Experiments in General, Organic, and Biological Chemistry

Experiments in General, Organic, and Biological Chemistry

Arne Langsjoen
Gustavus Adolphus College

Grover W. Everett
University of Kansas

Paul Lieder
E. I. Dupont De Nemours and Co.

Alfred J. Lata
University of Kansas

Harcourt Brace Jovanovich, Publishers
and its subsidiary, Academic Press
San Diego New York Chicago Austin Washington, D.C.
London Sydney Tokyo Toronto

ISBN: 0-15-525901-6

Printed in the United States of America

Cover design: Richard Kharibian

Preface

This manual is intended for students who have not necessarily had any previous background in chemistry and who are preparing for a career in nursing or other allied health occupations. The first part of the manual emphasizes *general chemistry* and is at the appropriate level for students in a one-semester liberal arts and sciences course in chemistry. In fact, the first ten experiments have been used for such a course at the University of Kansas for several years.

The manual can be used with any textbook written for students planning careers in the health sciences. There are more than enough experiments for a two-semester or a three-quarter course, thereby allowing flexibility in the laboratory curriculum. The general chemistry experiments (1–12) parallel the sequence of topics normally covered in an introductory course. These are followed by a series of experiments dealing with organic compounds, demonstrating a variety of organic reactions and laboratory techniques. Experiments in the last third of the manual illustrate the properties and methods of separation and analysis of selected biological compounds.

Each experiment is preceded by a statement of the *objectives* of the experiment and by *prelaboratory questions* designed to encourage the student to study the manual prior to the laboratory period. In the earlier experiments, answers to these questions are in the *discussion* section. More challenging questions are given at the end of each experiment. Special safety precautions are inserted, where appropriate, in the text of the *experimental procedure*.

Most of the experiments can be completed easily within a three-hour laboratory period. Students may work individually or in pairs, depending on the availability of equipment. The data sheets are designed to be separated from the manual and turned in to the instructor at the end of the laboratory period. Alternatively, some laboratory instructors may require that laboratory notebooks be kept by the students.

An Instructor's Guide for this manual is available that (1) lists equipment needed for each experiment, (2) gives directions for making the required reagents, (3) suggests topics for prelaboratory lectures, (4) discusses safety precautions, (5) describes ways in which laboratory time may be used most efficiently, and (6) provides answers to the prelaboratory and postlaboratory questions.

Preface

Contents

To the Student

This may be your first experience with the science known as chemistry, and you may have special reservations about the laboratory part of the course. No doubt you have heard tales of bad smells, fires, poisons, and even explosions in chemistry laboratories. Rest assured that the experiments in this manual have been carried out by thousands of students like yourself without a serious accident. These experiments were chosen so as to exclude chemicals and procedures that are truly dangerous. Whenever a potentially harmful process or chemical must be used, a CAUTION statement is given in the instructions. Use of common sense and strict adherence to the safety rules should minimize even minor accidents.

This laboratory manual was designed for students who have not necessarily had any previous experience with chemistry and who do not intend to become professional chemists. The first experiments provide an easy introduction to the equipment and techniques used in chemical laboratories. Those in the second half of the manual provide a good practical background for students who plan to enter one of the health care professions.

The laboratory is a place to learn and to solve problems. In many cases you will probably begin an experiment without a complete understanding of what you should do and what the experiment means. Don't be worried that you do not understand everything when you begin an experiment. After all, you are doing the experiment and taking the course to *learn* chemistry. Your efforts in the laboratory should be directed toward first understanding what is happening and then incorporating this knowledge into your overall understanding of chemistry. Ask for assistance from your laboratory instructor or your neighbor when you need it.

It is important that you read over an experiment, particularly the DISCUSSION section, *prior* to your laboratory period. You should be able to answer the PRELABORATORY QUESTIONS before you enter the laboratory. In some cases you may want to consult your chemistry textbook for a more detailed discussion.

We believe these experiments will help you understand chemistry, and we sincerely hope that you will find the laboratory part of your course to be both meaningful and fun!

To the Student

Safety Rules and First Aid Procedures

Telephone number for Emergency Medical Aid ______________________

Telephone number of Fire Department ______________________

Location of nearest:

Safety Shower ______________________

Eyewash Fountain ______________________

Fire Extinguisher ______________________

Fire Blanket ______________________

Emergency Exit ______________________

First Aid Station ______________________

GENERAL SAFETY REGULATIONS

1. *Approved* safety goggles must be worn at all times in the laboratory, except when their removal is authorized by the laboratory instructor.
2. Laboratory work will be performed only under the supervision of the laboratory instructor.
3. Unauthorized experiments are forbidden. When in doubt about the safety of an experiment, consult your instructor.
4. Horseplay and throwing objects in the laboratory can endanger others and are forbidden.
5. Know how to use the safety shower, eyewash fountain, fire extinguisher, and fire blanket.
6. Immediately report all injuries or accidents such as cuts, burns,

substances in the eye, chemical spills, and broken glass to your laboratory instructor.

7. Eating, drinking, and smoking in the laboratory are forbidden. Never taste or deeply inhale any laboratory chemical.
8. Dress appropriately. Loose-fitting clothing and long hair may fall into a burner flame. Do not use hair spray or other hair-styling chemicals prior to entering the laboratory. Shoes must be worn. Sandals and shorts are not permitted.
9. Do not point heated test tubes at anyone. Take care when transporting long or sharp items of equipment so as not to injure anyone.
10. Always use a pipet bulb with a pipet. Never draw a liquid into a pipet by mouth.
11. Use equipment properly. Books, beakers, and boxes are not suitable as supports for laboratory apparatus.
12. Be cautious when testing odors of chemicals. Use your hand to waft vapors toward your nose. Use hoods when handling volatile chemicals. Prior to beginning an experiment, familiarize yourself with properties of the chemicals to be used.
13. Keep your laboratory drawer and bench clean and make sure all equipment is clean before you leave the laboratory. Clean up any chemical spills immediately.
14. Consult your laboratory instructor before discarding chemicals. Special methods of disposal are necessary for certain chemicals.
15. It is a good idea to wash your hands frequently when handling chemicals. Wash your hands when leaving the laboratory.

FIRST AID

Be prepared to administer first aid as follows:

Chemicals in the Eyes

Immediately flush with copious amounts of water at the eyewash fountain and continue to wash for *at least 15 minutes*. The injured person should keep his or her eyes open, or hold them open with the eyelid rolled back, while flushing with water.

If the injured person is wearing contact lenses, remove them *after* flushing the eyes with water for several minutes, then continue flushing.

Finally, cover the eyes loosely with a clean bandage or cloth and see a doctor. Emergency medical treatment and an ambulance may be desirable.

Corrosive Chemicals on the Skin

Wash immediately with lots of cold water. Remove any clothing that has been contaminated with the chemical and continue to wash the skin with cold water for at least 15 minutes. If the skin is not broken, wash the area gently with a mild detergent and water, then proceed immediately to a doctor if the burn appears serious. Special treatments for specific chemicals will be suggested as they are encountered in the experiments. *Do not* use neutralizing chemicals, ointments, creams, or lotions. If in doubt about the seriousness of the burn, see a doctor. Emergency medical treatment and an ambulance may be desirable.

Burns Due to Fires and Hot Objects

Hold the burned area under cold water for several minutes, then wrap with sterile gauze. *Do not* apply ointment to the burn. If the burn appears serious, see a doctor.

Cuts

For serious bleeding, call for medical help. Try to reduce the bleeding by applying direct pressure to the wound and, if possible, elevate the injury above the heart. *Do not* apply a tourniquet.

If the cut is minor, allow it to bleed for several seconds, then wash it well with cold water and remove any visible glass or other foreign objects. Apply disinfectant and bandage and go to the Health Service for a more thorough check.

Fires

If clothing or hair catches fire, either smother the fire with a fire blanket (or any immediately available heavy clothing such as a coat) *or* drench the victim with the safety shower, whichever is nearer. The spread of flames will be minimized if the victim drops to a horizontal position. Treat burns as described above.

If the fire is confined to a reaction apparatus, extinguish it with a fire extinguisher. A small fire in a beaker can usually be extinguished by covering it with a wet towel—but do not attempt to move the burning apparatus to another location.

In case of larger fires, evacuate the room and call the fire department.

Ingestion of Chemicals

Call for medical help. If the victim is conscious, have him or her drink large amounts of water. Try to determine what was ingested in order to inform medical personnel.

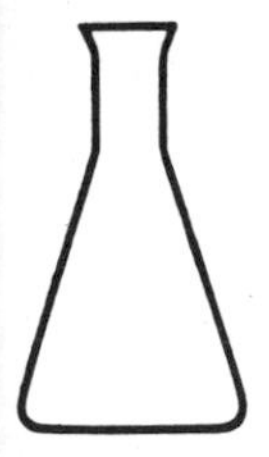

Laboratory Equipment

Items of laboratory equipment used during the experiments in this manual are illustrated on the following pages. Special items that are needed for some of the experiments will be identified by your instructor.

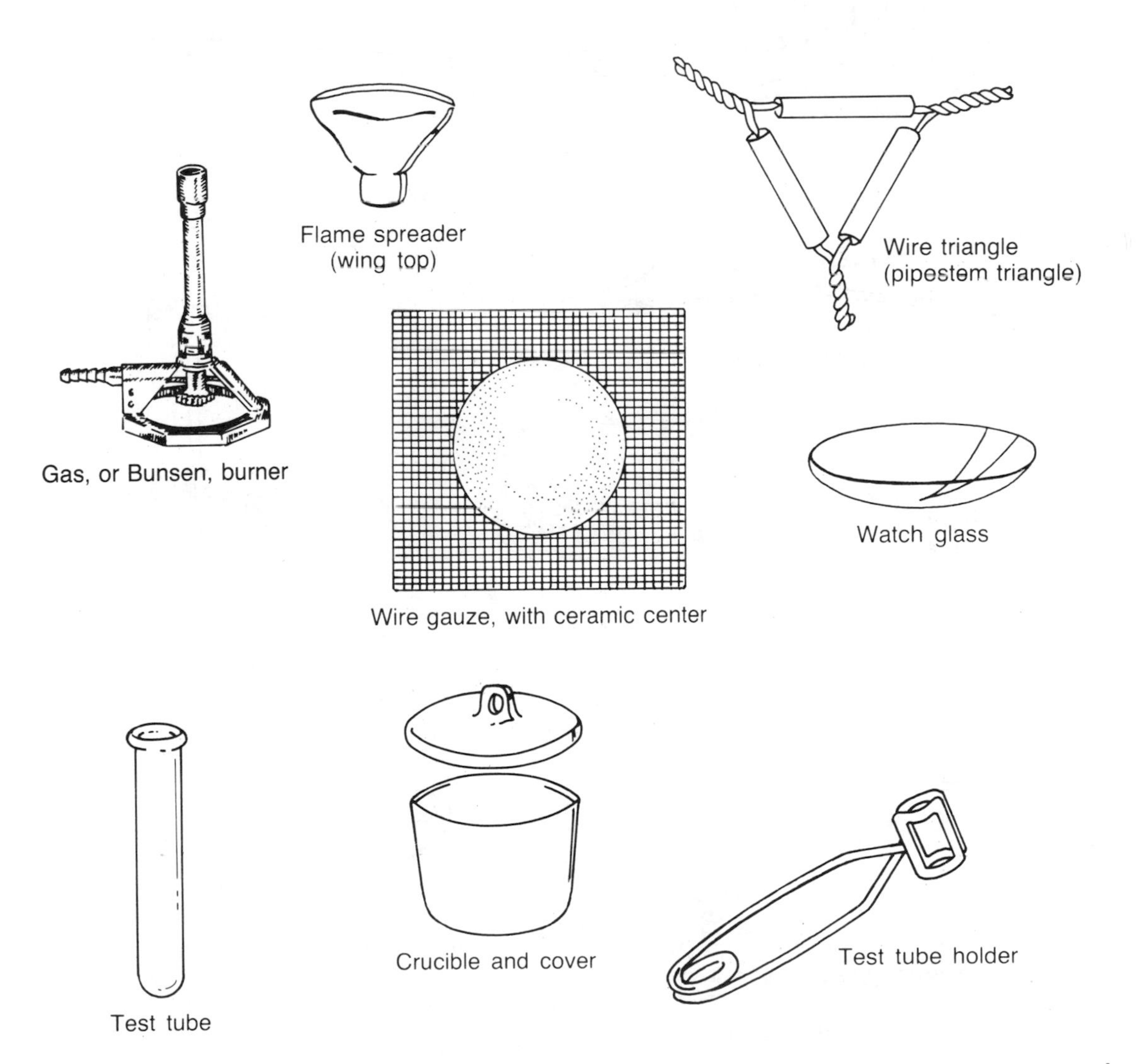

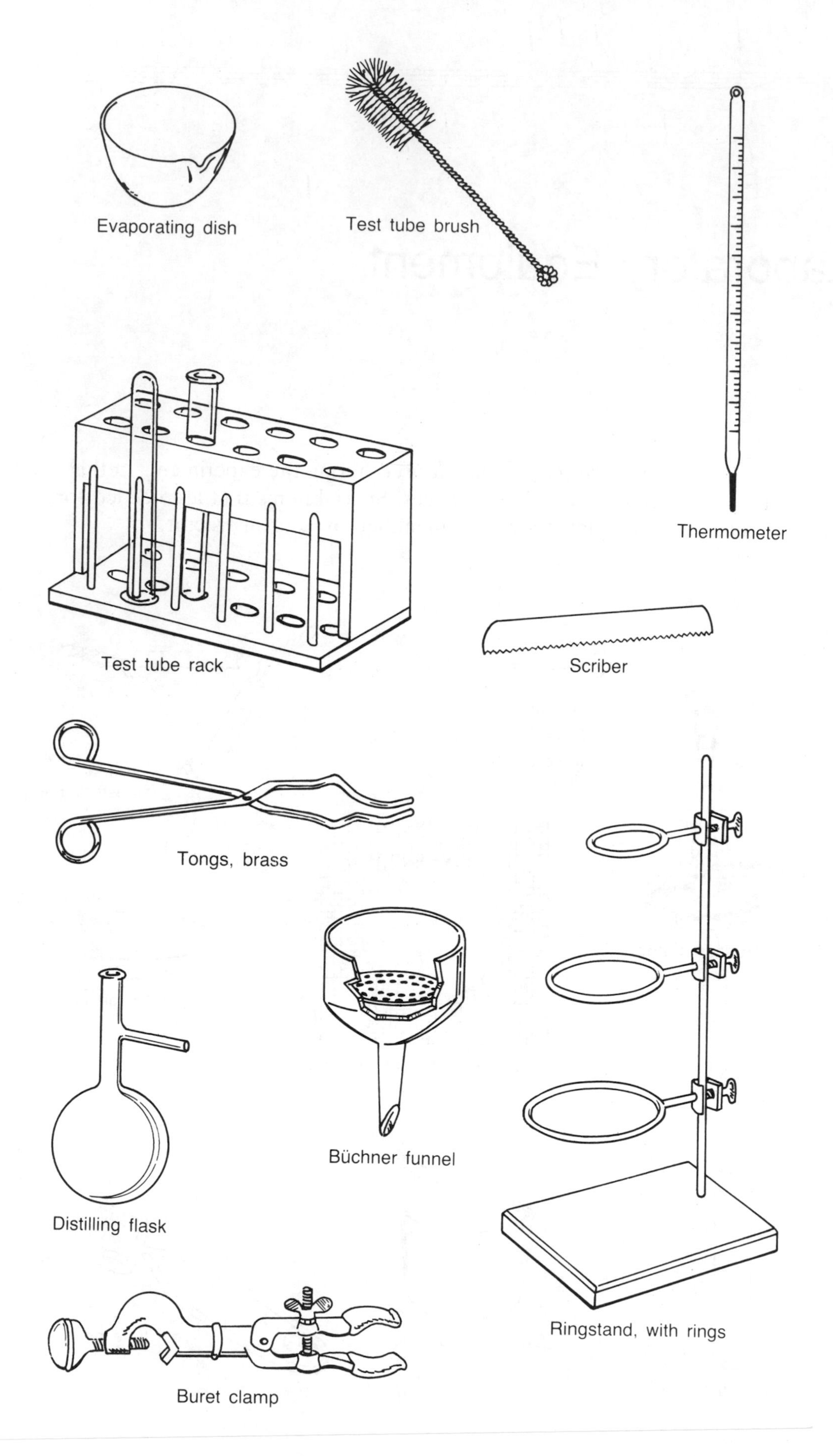
Evaporating dish
Test tube brush
Thermometer
Test tube rack
Scriber
Tongs, brass
Distilling flask
Büchner funnel
Ringstand, with rings
Buret clamp

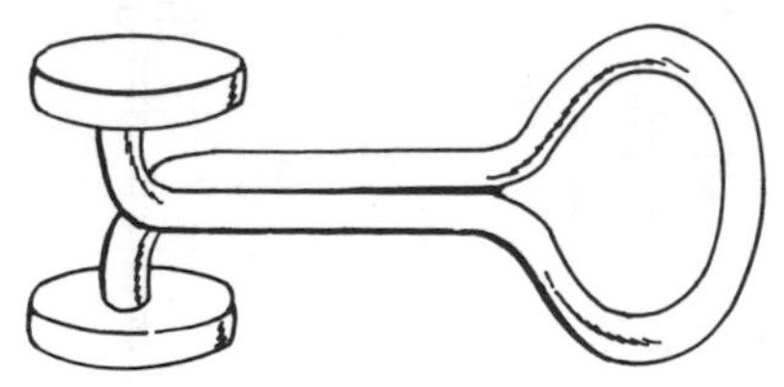
Pinchcock clamp, Hoffman

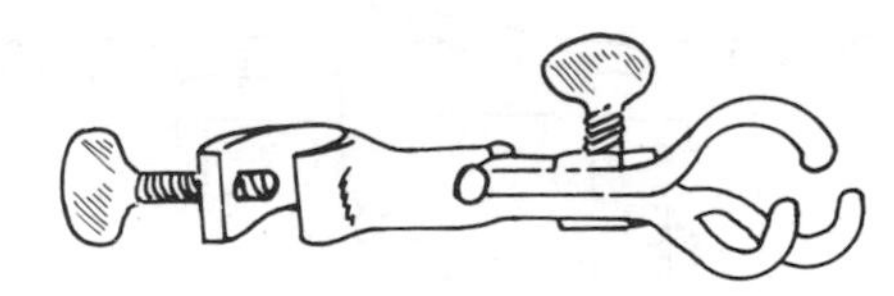
Condenser clamp

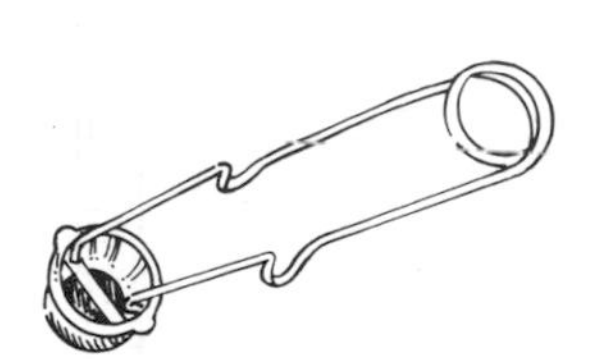
Gas lighter

Deflagration spoon

Wash bottle

Forceps

Triangular file

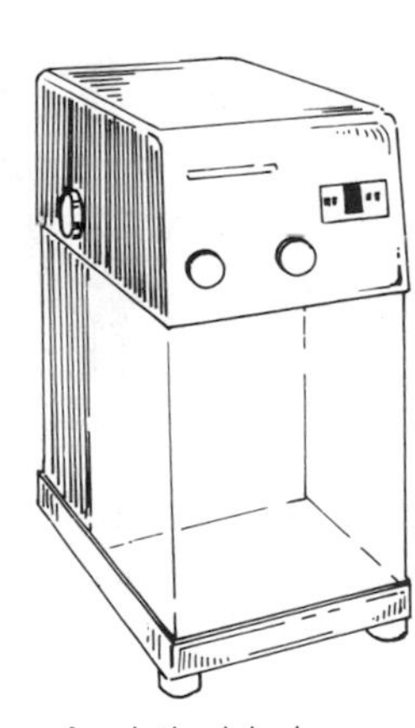
Analytical balance

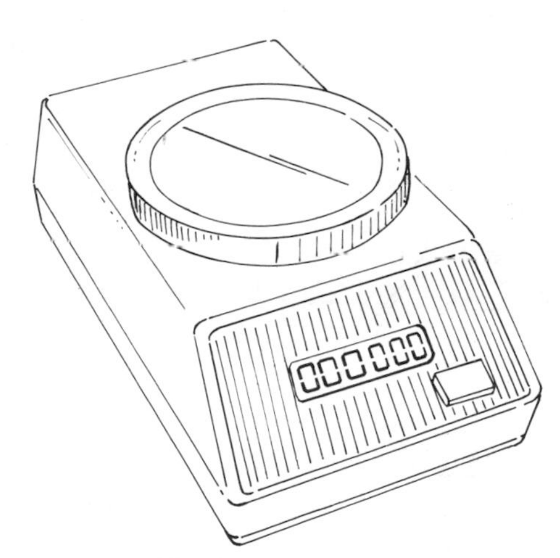

Top-loader balance

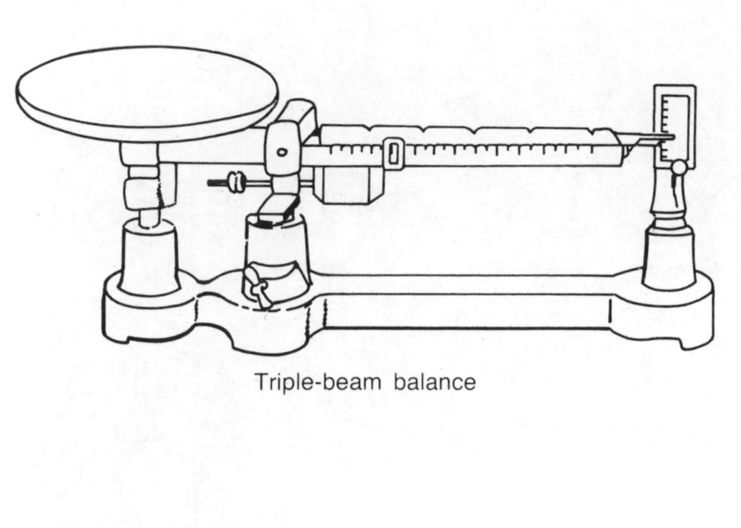

Triple-beam balance

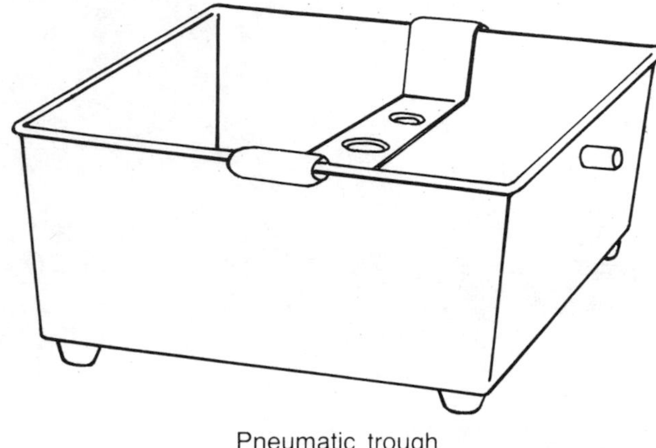
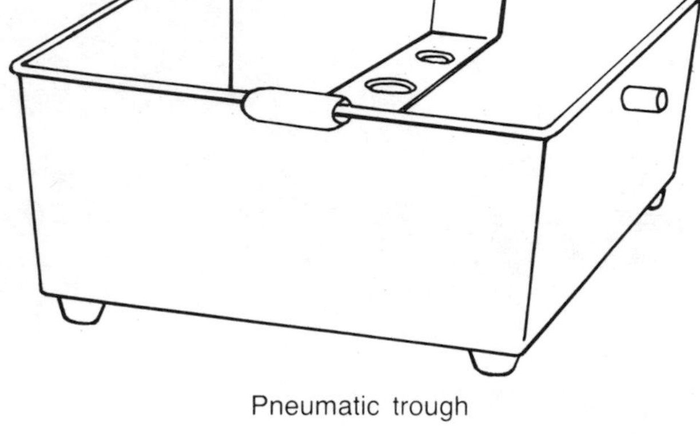

Pneumatic trough

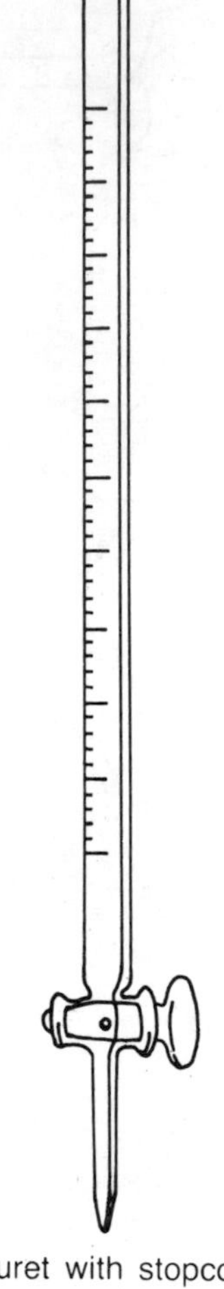

Buret with stopcock

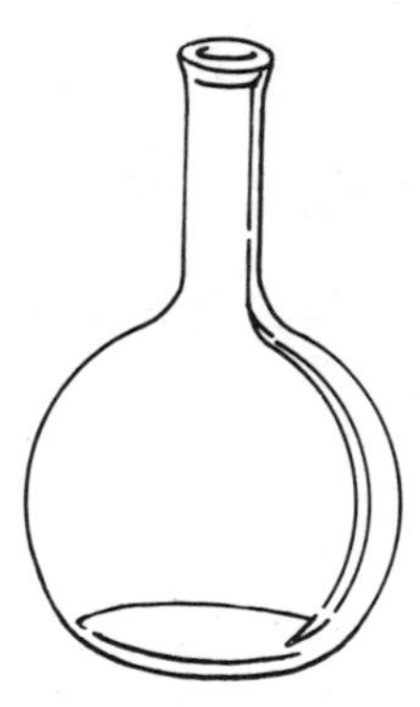

Flask, Florence

Beaker

Funnel

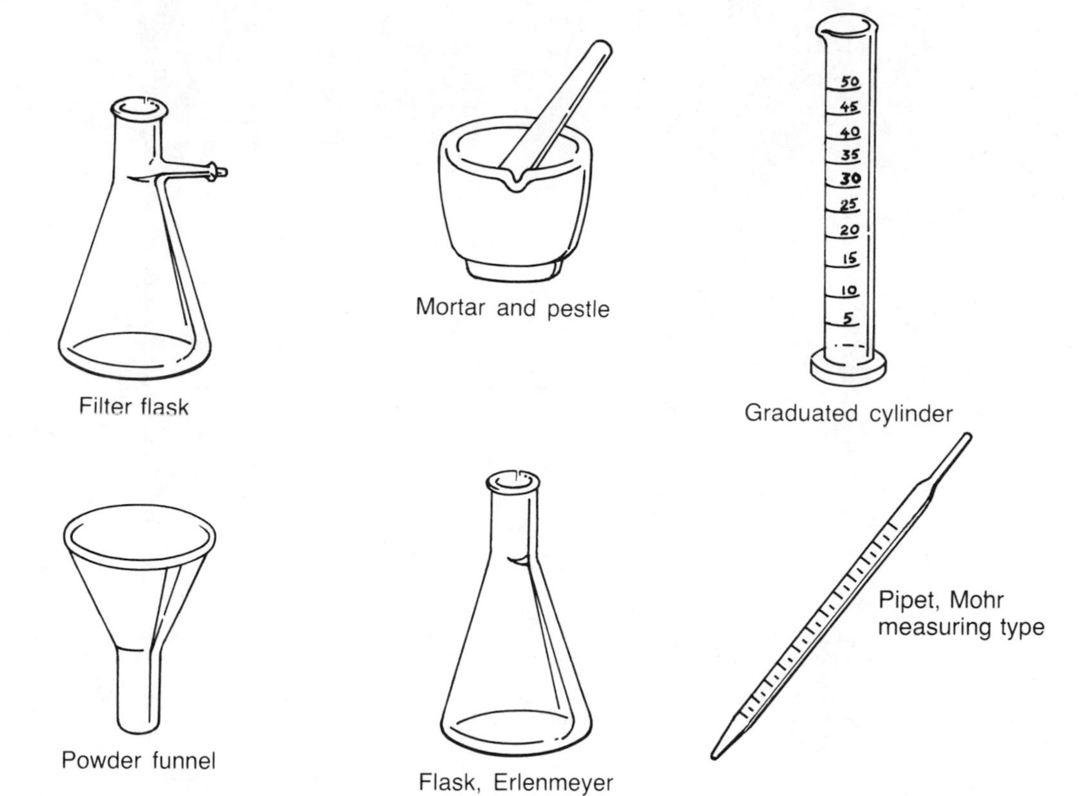

Filter flask
Mortar and pestle
50
45
40
35
30
25
20
15
10
5
Graduated cylinder
Powder funnel
Flask, Erlenmeyer
Pipet, Mohr
measuring type

EXPERIMENT 1

Introduction to the Laboratory

OBJECTIVES

1. To become acquainted with the use and operation of two types of laboratory balances.
2. To learn how to light and adjust a gas burner.

PRELABORATORY QUESTIONS

1. Why should chemicals never be placed directly on the pan of a balance?
2. Explain the procedure for determining the mass of a substance "by difference."
3. Describe where the hottest part of a gas burner flame may be found.

DISCUSSION

Chemistry is an *experimental* science. The chemical laboratory is a place where scientific theories are born and tested. The purpose of the laboratory part of a chemistry course is to enable you to gain first-hand experience with laboratory equipment, chemicals, experimental procedures, and problem-solving methods. This experience should help you to understand how a chemist thinks and acts.

During this course, you will very likely begin experiments without a complete understanding of how to perform the experiment or of the scientific principles involved. The laboratory period is intended to be a *learning* experience. Your laboratory instructor will answer questions you may have during the course of the experiment.

It is important for you to be able to determine the mass of a substance quickly and accurately. (This procedure is often called "weighing.") Masses may be determined using a *balance*. During this experiment you will learn to use two types of balances.

Top-loader or *open-beam* balances are designed for obtaining approximate masses of chemicals to be used in reactions and for determining masses in other situations where extreme accuracy is *not* important. Masses can be read to the nearest 0.1 g or 0.01 g on most top-loader or open-beam balances.

An *analytical* balance is a precision instrument capable of determining masses to the nearest 0.1 *milli*gram (0.0001 g). It is used whenever very accurate masses are needed for quantitative calculations.

Chemicals must *never* be placed directly on the pan of a balance. Not only would this corrode the balance pan, but also it would be difficult to collect *all* of the chemical after its mass has been determined. Furthermore, the pan would have to be cleaned after each use. Thus you should always use a receptacle, such as a piece of filter paper, weighing paper, a sample vial, or a small beaker, that may safely be placed on the balance pan to contain the chemical. First the mass of the empty container, M_1, is determined, then the combined mass of the container and the chemical substance, M_2, is determined. The mass of the chemical substance alone is the *difference* in these masses, $M_2 - M_1$. This is called "weighing by difference."

Your instructor will demonstrate how to operate both types of balances, then you will practice using them by determining masses of two small items of laboratory equipment.

Since its invention in 1855, the *Bunsen burner* and its various modifications have been used in laboratories throughout the world as convenient sources of intense heat. These burners generally use natural gas (methane) as a fuel, because this is available in most laboratories. The gas is mixed with air that enters the burner through holes near the base of the barrel. The flame becomes hotter as more air is introduced. The most intense heat occurs just above the *blue* portion of the flame. This appears inside the larger, violet part of the flame. You will learn how to light and adjust a gas burner, and then you will use it to soften glass tubing.

MATERIALS NEEDED

top-loader (or open-beam) balance
analytical balance
weighing paper
burner wing top
crucible
gas burner
triangular file (or scriber)
soft glass tubing, 6 mm × 30 cm

EXPERIMENTAL PROCEDURE

A. Operation of Balances

1. Place a piece of weighing paper on the pan of a top-loader (or open-beam) balance. Determine its mass to the nearest 0.1 g (or 0.01 g,

depending on the make of the balance), and record the mass (M_1) on the data sheet. Then place an empty crucible (without cover) on the weighing paper and record the combined mass of crucible and weighing paper on the data sheet (M_2). The *difference* in the two masses is the mass of the crucible alone.

2. Repeat the above experiment, using a burner wing top instead of the crucible. Record M_1 and M_2 on the data sheet and calculate the mass of the wing top.
3. Place a piece of weighing paper on the pan of an *analytical* balance and determine the mass of the paper to the nearest 0.0001 g. (All masses determined on the analytical balance should be read to the fourth decimal place.) Record this mass, M_1, on the data sheet. Then place the crucible on the weighing paper; determine the combined mass, M_2, and record this mass on the data sheet. Calculate the mass of the crucible.
4. Repeat procedure 3, using the burner wing top instead of the crucible. Record M_1 and M_2 on the data sheet and calculate the mass of the wing top.

B. Use of the Gas Burner

1. Examine your gas burner (see Figure 1.1). Find the *air intake holes* near the base of the barrel. These may be opened or closed by rotating either the barrel or a sleeve around the barrel. Locate the *gas flow knob* on the underside of the burner. This is used to control the size of the flame. Your instructor will demonstrate how to connect a burner to the laboratory gas outlet.
2. Close the gas flow valve by rotating the knob clockwise (as viewed from the bottom) as far as it will turn, then open it about one-half turn. Close the air intake holes as much as possible. Place the burner on the bench top and make sure no flammable objects are directly above it. Open the valve from the laboratory gas supply to its *maximum* position. Light the burner and adjust the gas flow valve beneath the burner so as to produce an orange-yellow flame about 10 cm high.

   The size of the flame is always controlled by adjusting the gas flow valve beneath the burner and *not* the valve at the laboratory gas outlet (this should be fully open).

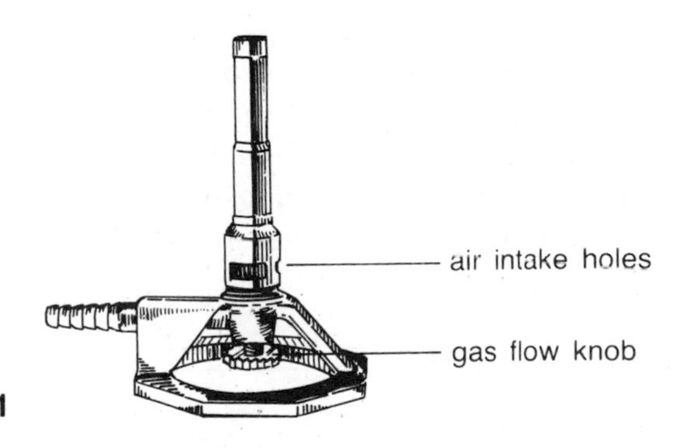

Figure 1.1

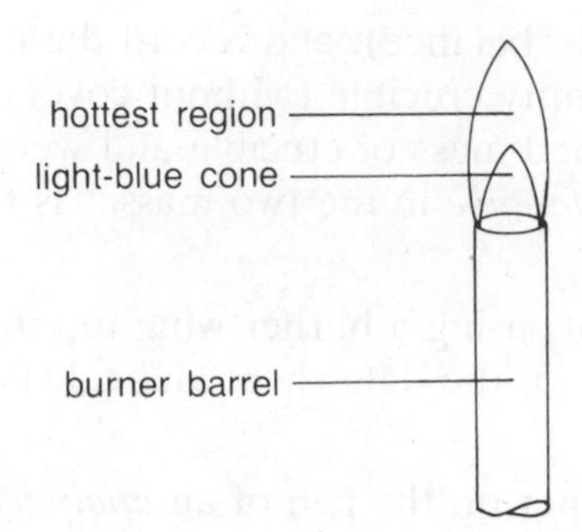

Figure 1.2

3. Slowly open the air intake holes. The flame will develop a violet color, and a light-blue inner cone will appear (see Figure 1.2). Increase the air supply until the light-blue cone burns vigorously. If the flame goes out, decrease the air intake and relight the burner. Adjust the gas flow valve so that the light-blue portion of the flame is about 5 cm high. The flame may now be used for softening glass tubing.

C. Heating Glass Tubing

When heating a glass tube, rotate the tube constantly to heat all sides equally. This is especially important prior to *bending* the tube, because if the tube is not heated uniformly, the inner bore of the tube will close when the tube is bent.

CAUTION: Hot glass has the same appearance as cool glass. A piece of glass that is cool at one end might be quite hot at the other end! Allow several minutes for hot glass to cool before picking it up.

1. Cutting and Polishing Your instructor will give you a piece of glass tubing approximately 30 cm in length. Cut this into two sections in the following manner: Use a small triangular file to make a scratch in the surface about halfway along the tube. Wet the scratch with a drop of water. Using both hands, hold the tube in a cloth towel with the scratch pointed away from you and bend the tube gently away from you. The tube should break easily at the scratch. If not, make the scratch deeper.

Your instructor will demonstrate how to round off or "fire polish" the sharp ends of the tubes using a gas burner. After you polish both ends of the tubes, allow several minutes for them to cool before proceeding to part 2.

2. Making Pipets Hold a piece of polished tubing in both hands and, while rotating the tube continuously, heat a section near the center. Hold the tube just above the inner, light-blue part of the flame. When the glass feels fairly soft, remove it from the flame and immediately pull the ends apart to stretch and narrow the heated section. Do not overheat the tube or stretch it too far. Your tube should look similar to the one in Figure 1.3.

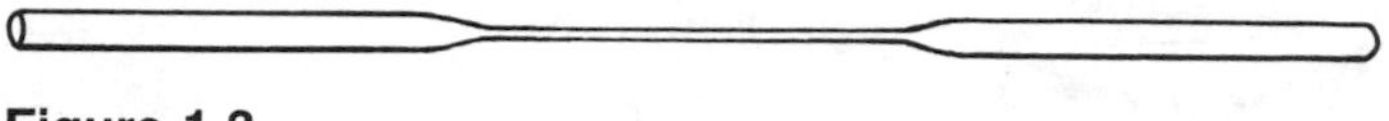

Figure 1.3

After the tube has cooled, *carefully* make a scratch with the file at the center of the narrow section and break the tube into two dropping pipets.

3. Making a Bend Turn off the burner by closing the valve at the laboratory gas supply. Allow the burner to cool. Attach a wing top to the burner barrel and relight the burner. The wing top spreads the flame so that a larger section of glass can be heated. Adjust the air intake and gas flow controls so that the light-blue part of the flame is about 2 cm high.

Heat the center portion of the remaining glass tube while rotating the tube continuously in the flame. Your tube should be held *parallel* to the spread flame and just above the light-blue portion. When the glass is quite soft, remove it from the flame and make a right-angle bend before the glass cools. The center bore of the tube should remain open after bending (see Figure 1.4).

Show your pipets and bent tube to the instructor for approval.

Figure 1.4

DATA SHEET

Name: ______________________ Lab Section: ______ Date: ____________

INTRODUCTION TO THE LABORATORY

A. Operation of Balances

	Top-loader	Analytical
Mass of crucible and paper, M_2	________	________
Mass of paper, M_1	________	________
Mass of crucible, $M_2 - M_1$	________	________
Mass of wing top and paper, M_2	________	________
Mass of paper, M_1	________	________
Mass of wing top, $M_2 - M_1$	________	________

C. Heating Glass Tubing

QUESTIONS

1. What should a student do in case of an accident in the laboratory—even a minor accident?

2. Predict the most frequent accident that occurs when students are softening glass for the first time.

3. Compare the masses of the crucible obtained from the two types of balances. Round off the mass obtained on the analytical balance to the nearest 0.1 g (or 0.01 g if your top-loader balance has this accuracy). Should this rounded-off mass be identical to the mass obtained with the top-loader balance? Explain your answer.

4. An object is weighed twice on the same analytical balance, once using a beaker as the receptacle and once using a piece of paper as the receptacle. Should the two masses obtained for the object be identical? Explain.

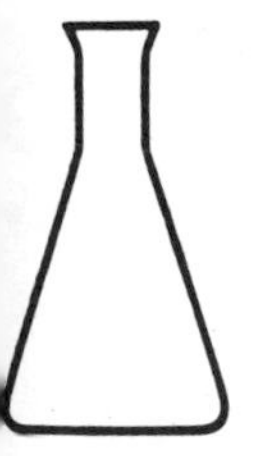

Chemical and Physical Properties

OBJECTIVES

1. To introduce the concepts of physical and chemical properties and physical and chemical changes.
2. To provide first-hand experience with physical and chemical changes.

PRELABORATORY QUESTIONS

1. Explain what happens when a substance undergoes a chemical change.
2. Name three well-known physical properties of water.
3. Does a chemical change or a physical change occur when water freezes?

DISCUSSION

Chemistry is the branch of science concerned with the *composition* of matter and methods by which *changes in composition* can be carried out. That is, chemists are interested in the makeup of substances and in how one substance can be changed into another.

A *chemical property* of a substance is a property that is a measure of the tendency of that substance to react with other substances to undergo a change in composition, i.e., a *chemical change*. For example, many substances react with oxygen (a process called burning or combustion) to form new compounds called oxides. Also, the element sodium reacts with the element chlorine to form the compound sodium chloride (table salt). These are chemical changes. The inherent *ability* of sodium to react with chlorine and the ability of many substances to react with oxygen are examples of chemical properties. Chemical properties may be used to identify a substance.

Substances may also be identified by their *physical properties*. Physical properties are those that are independent of chemical reactivity. Examples are color, density, hardness, melting temperature, boiling temperature, and solubility. Physical properties are associated with *physical changes*, which are changes in the physical appearance of a substance but which involve no change in composition. For example, when a cube of ice melts, the liquid water formed has the same chemical composition as the frozen water in the ice cube. Therefore the process of melting is a physical change.

In this experiment you will examine some chemical and physical properties of the elements iron (Fe) and sulfur (S) and of a compound, iron sulfide (FeS), which is formed when these elements react with each other.

MATERIALS NEEDED

- 150 mm test tube
- small test tubes
- test tube holder
- filter paper
- glass funnel
- ringstand with small ring
- 25 mL graduated cylinder
- 250 mL beaker
- deflagration spoon
- wood splint
- watch glasses
- iron pieces (rectangular cross section)
- sulfur pieces (2–3 g)
- powdered iron
- powdered sulfur
- 6 M HCl
- lead acetate paper
- carbon disulfide (instructor only)
- metric ruler
- gas burner

EXPERIMENTAL PROCEDURE

Before proceeding, you should review the section on safety regulations and first aid procedures given at the beginning of this laboratory manual.

A. Density of Iron

Measure the three dimensions (a, b, c) of a piece of iron to the nearest 0.05 cm. Record these dimensions on the data sheet and calculate the volume of the piece of iron. Determine the mass of the piece of iron to the nearest 0.1 g using a top-loader or open-beam balance. Remember to weigh by difference. Record the mass on the data sheet, then calculate the density of iron. Be sure to use the proper units in the calculated volume and density.

B. Density of Sulfur

Determine the mass of a piece of sulfur (by difference) to the nearest 0.1 g using a top-loader or open-beam balance. Record the mass on the data sheet.

The *volume* of an irregular piece of sulfur may be determined by measuring the volume of water it displaces when immersed. Fill a 25 mL graduated cylinder with water to a volume of approximately 20 mL. Read the volume of water to the nearest 0.1 mL and record this initial volume, V_i, on the data sheet. Immerse a piece of sulfur in the water as shown in Figure 2.1.

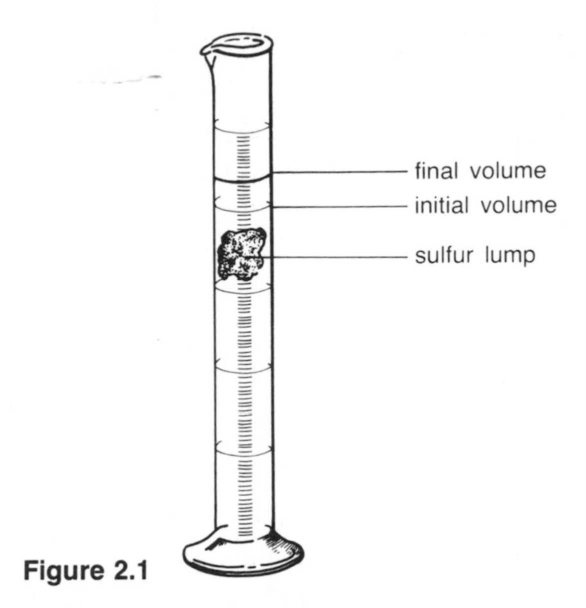

Figure 2.1

Read the final volume, V_f, to the nearest 0.1 mL and record it on the data sheet. The volume of the piece of sulfur is the same as the *difference* in these two volumes of water.

Use the measured mass and volume of the piece of sulfur to calculate its density and record the density on the data sheet.

C. Combustion of Sulfur

CAUTION: The sulfur dioxide gas formed in this part of the experiment can be irritating to the eyes, throat, and lungs if present in sufficient concentration. Therefore *use only the amount of sulfur specified below and perform the reaction under a fume hood.*

Fill a deflagration spoon no more than one-quarter full of powdered sulfur. Light a gas burner and heat the sulfur *under a fume hood* until it burns. Observe the color of the sulfur flame and the odor of burning sulfur. Record your observations on the data sheet.

D. Physical Changes in Sulfur

Set up a ringstand, ring, and glass funnel (under a fume hood if possible) as illustrated in Figure 2.2. Fold a piece of filter paper (as shown by your instructor) and place it in the funnel. Also secure a 250 mL beaker and fill it half full with water. Fill a 150 mm test tube approximately half full of powdered sulfur. Light a gas burner and adjust the air intake holes so that a cool (yellow) flame is produced.

1. *Liquid Sulfur.* **CAUTION:** *Molten sulfur is very hot!* Using a test tube holder, heat the test tube of sulfur from *top to bottom* gently and slowly under a hood. Make sure the open end of the test tube is not pointed at anyone. If the sulfur ignites at the mouth of the test tube, remove the test tube from the heat and blow out the flame. Do not overheat the sulfur, because it will develop a dark color and become difficult to pour. When

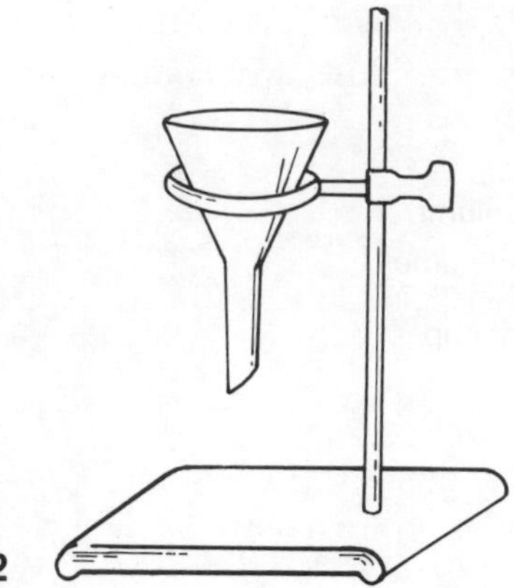

Figure 2.2

the sulfur melts, pour *half* of it into the beaker of water and *half* onto the filter paper. Briefly describe liquid sulfur on the data sheet.

2. *Monoclinic Sulfur*. As soon as the surface of the molten sulfur in the filter paper develops a crust, remove the paper from the funnel and open it flat. *Be careful*: The sulfur is still very hot! Examine this sulfur carefully and describe its appearance on the data sheet. The tiny fibers are crystals of the monoclinic form of sulfur.

3. *Plastic Sulfur*. Remove the sulfur from the beaker of water. Try to bend and twist the sulfur. Record your observations on the data sheet. This noncrystalline form of sulfur is called "plastic" sulfur.

E. Formation of Iron Sulfide

Use a top-loader or open-beam balance to measure *approximately* 2 g each of powdered iron and powdered sulfur. Use pieces of paper to contain these elements—do not put the chemicals directly on the balance pan! Combine the iron and sulfur on one of the pieces of paper and mix them well with a wood splint. Pour the mixture into a small test tube.

Hold the test tube with a test tube holder and heat the mixture gently over a gas burner until you see a *reaction* start, as evidenced by a glow in the test tube. Remove the tube from the flame. It may take several heatings for the entire sample to react. Heating iron and sulfur causes a chemical reaction to occur between these elements to form the *compound* iron(II) sulfide, FeS.

Allow the test tube to cool thoroughly. Then wrap the test tube in a towel and hit it carefully with an iron ring to break the tube. Examine the product and describe its appearance on the data sheet.

F. Solubility in Carbon Disulfide (Instructor demonstration)

CAUTION: Since carbon disulfide is extremely flammable and toxic, this part of the experiment will be *demonstrated by your laboratory instructor* under a fume hood. *All flames in the laboratory must be extinguished during this demonstration*. You should observe the demonstration carefully and record your observations on the data sheet.

1. *Addition of Carbon Disulfide*. Place small samples of powdered iron, powdered sulfur, and iron sulfide (from the reaction in step E) in

separate small test tubes. Add approximately 2 mL of carbon disulfide to each tube and swirl the tube gently. Note any changes and record your observations on the data sheet.

2. *Evaporation of Carbon Disulfide.* Pour the carbon disulfide from the test tubes onto three separate watch glasses under a fume hood, leaving any solid material in the test tubes (this is called decanting). Allow the carbon disulfide to evaporate, then examine the watch glasses. What do you observe?

G. Test for Sulfide Ion

Lead acetate paper may be used to detect the presence of sulfide ion, S^{2-}, in a sample to which acid is added. If sulfide ion is present, the acid reacts with the sample to form hydrogen sulfide gas (a chemical change!), which will darken the lead acetate paper.

Place small samples of Fe, S, and FeS in separate test tubes and add approximately 2 mL of 6 molar hydrochloric acid (6 M HCl) to each. Hold a piece of moist lead acetate paper at the opening of each test tube. Record your observations.

DATA SHEET

Name: ______________________ Lab Section: ______ Date: __________

CHEMICAL AND PHYSICAL PROPERTIES

A. Density of Iron

Dimensions of the sample a __________

b __________

c __________

Volume ($a \times b \times c$) __________

Mass of iron and weighing paper, M_2 __________

Mass of weighing paper, M_1 __________

Mass of iron, $M_2 - M_1$ __________

Density of iron (mass/volume) __________

B. Density of Sulfur

Mass of sulfur and paper, M_2 __________

Mass of paper, M_1 __________

Mass of sulfur, $M_2 - M_1$ __________

Final volume of water, V_f __________

Initial volume of water, V_i __________

Volume of sample ($V_f - V_i$) __________

Density of sulfur (mass/volume) __________

C. Combustion of Sulfur

D. Physical Changes in Sulfur

1. *Liquid Sulfur*

2. *Monoclinic Sulfur*

3. *Plastic Sulfur*

E. Formation of Iron Sulfide

F. Solubility in Carbon Disulfide (Instructor demonstration)

1. *Addition of Carbon Disulfide*

Fe

S

FeS

2. *Evaporation of Carbon Disulfide*

Fe

S

FeS

G. Test for Sulfide Ion

Fe

S

FeS

QUESTIONS

1. From your observations, list as many *physical* properties as you can for:

   Fe

   S

   FeS

2. From your observations, list as many *chemical* properties as you can for:

   Fe

   S

   FeS

Separation of Components of Mixtures

OBJECTIVES

1. To demonstrate how substances in a mixture may be separated by utilizing differences in their physical properties.
2. To gain first-hand experience with the commonly used laboratory separation techniques of chromatography, distillation, and extraction.

PRELABORATORY QUESTIONS

1. What physical properties are utilized in the four processes: extraction, distillation, flotation, and chromatography?
2. During the separation of mixtures of liquids by distillation, changes in the boiling temperature of the mixture occur. Explain why this happens.

DISCUSSION

Elements and compounds are classified as "pure substances" because throughout a sample the composition is uniform, and samples from different sources have the same composition. Industrialized nations require large quantities of pure substances for manufacturing purposes. However, most materials found in nature are *mixtures* of two or more pure substances. Therefore, economical methods of separating components of mixtures must be employed. Many of these methods take advantage of differences in *physical properties* of the substances in a mixture. Let us examine some physical properties with which you are familiar and explain how these might be used to separate components of a mixture.

Boiling point: If the components of a mixture are liquids with relatively low boiling points (0–300°C), it may be possible to separate them by a process called *distillation*. In this process, the mixture is heated carefully so that its temperature increases slowly. When the boiling point of one of the components is reached, the mixture will boil while that component is converted to the vapor phase. The vapor is conducted through a "condenser," a tube which is maintained at a cool temperature so that the vapor condenses back to a liquid. This liquid is one of the pure substances from the mixture.

After the first component of the mixture has thus been removed, the temperature of the mixture increases until the boiling point of another component is reached. This component is also removed from the mixture by vaporization and condensation. This process continues until all the components of the mixture are separated and collected. Each component may be identified during the distillation by its characteristic boiling point. In actual practice, using the simple equipment available in a general chemistry laboratory, the distillation does not result in *complete* separation of the components. Traces of a lower-boiling component usually remain in the flask when a higher-boiling component begins to distill.

Solubility: Different substances often have quite different solubilities in a particular solvent, and this physical property can be utilized to separate the substances. For example, if two compounds are present in a mixture, we can "extract" the mixture with a solvent that dissolves only one of the compounds. *Extraction* is a process whereby the mixture and solvent are combined and mixed thoroughly. The soluble substance dissolves in the solvent, leaving the insoluble substance as a residue that can be removed by filtering. The soluble substance can then be recovered by evaporating the solvent.

Density: If the components of a solid mixture have different densities, it may be possible to separate one or more of them by *flotation* in a liquid. The low-density components rise (float) to the surface of the liquid, while those of higher density sink. The separate components may then be collected by careful pouring or by filtering.

Molecular attraction to a solid surface: Due to differences in their molecular structures, substances in a mixture frequently differ in the extent to which they are attracted to the surface of a material such as paper. This physical property is widely used for separating and identifying components of mixtures, particularly when only small quantities of the mixtures are available.

In the technique known as *paper chromatography*, a solvent flows slowly along the surface of a piece of paper on which a mixture has been deposited. The components of the mixture move with the solvent flow at different rates that depend on their relative abilities to "stick" to the paper.

Thus, if solvent is passed over the surface of the paper for a sufficient length of time, the components of the mixture will be separated from each other on the surface of the paper. After the paper is dried, the individual substances may be removed. In some cases, the chemical identity of a substance can be determined by measuring the distance it travels along the paper relative to the distance traveled by the solvent.

During this experiment you will separate the components of several mixtures using the methods discussed above.

MATERIALS NEEDED

- chromatography paper
- felt-tip pen
- pencil
- 200 mm test tube
- scissors
- ice
- 125 mL distilling flask
- condenser with adapter
- rubber tubing
- thermometer
- ringstands (2) and small ring
- clamps
- 50 mL beakers (2)
- 250 mL beaker
- heating mantle
- powerstat transformer
- powdered sulfur
- NaCl
- sand
- mixture of sand, S, NaCl
- cork stoppers
- 25 mL graduated cylinder
- boiling chips
- "unknown" liquid mixtures
- aluminum foil
- funnel
- filter paper
- ruler

EXPERIMENTAL PROCEDURE

A. Paper Chromatography

1. Fold a piece of chromatography paper in half lengthwise. Use scissors to cut one end, as shown in Figure 3.1, so that the paper is pointed at the fold. With the felt-tip pen, make a small spot at the center of each half of the folded paper about 4 cm up from the pointed bottom (Figure 3.1). Use a pencil to mark the position of the ink spots at the edge of the paper.
2. Fill a 200 mm test tube with water to a depth of about 2 cm. Dry the inside wall of the test tube above the liquid level with a tissue. Insert the chromatography paper into the test tube so that the pointed end is below the water level. Place the test tube in an upright position and allow the

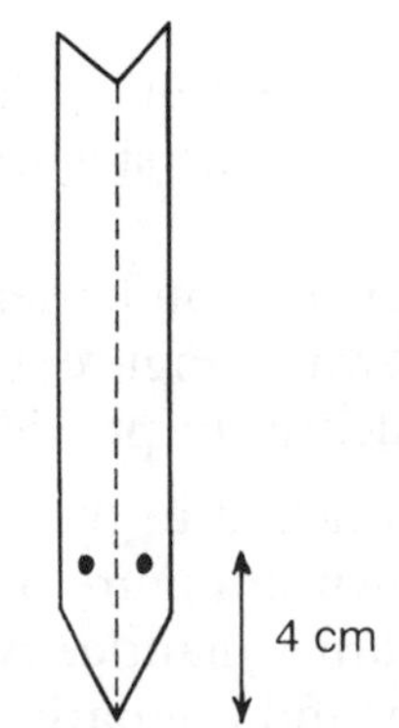

Figure 3.1

water to migrate up the paper until the leading edge of the water is about 2 cm from the top of the paper. Remove the paper and measure the distance (in cm) from the pencil mark to the leading edge of the water and record on the data sheet.

3. Allow the paper to dry. Measure the distances from the pencil mark to the leading edges of each color and record both the color and the distances on the data sheet. Calculate the ratio of these distances to the distance traveled by the water. Tape or staple the paper to the data sheet.

B. Distillation

Your instructor will give you a mixture of two liquids without revealing their identities. In this part of the experiment you will separate the mixture into its components by distillation and attempt to identify the components by their boiling temperatures.

CAUTION: The liquids used in this part of the experiment were selected for their low toxicities. However, *these liquids are very flammable. Do not use gas burners* for the distillation. Furthermore, *no open flames* of any kind should be present anywhere in the laboratory during this experiment.

1. Record the number of your "unknown" mixture for distillation on the data sheet. Place 25 mL of the mixture into a 125 mL side arm distilling flask and add a boiling chip.
2. Assemble the components of the distillation apparatus as shown in Figure 3.2:
   a. Attach the distilling flask to a ringstand with a clamp. Attach a small ring to the ringstand below the flask and use it to hold a heating mantle around the bottom of the flask. Attach the power cord from the heating mantle to the powerstat.
   b. Insert the thermometer and cork stopper into the top of the flask and adjust the height of the thermometer so that the *top* of the mercury bulb is at the level of the sidearm of the flask.
   c. Wet the inlet and outlet nipples on the condenser (for lubrication), then connect rubber tubing to these. Use a clamp to attach the condenser to a second ringstand. Connect the hose from the *bottom* of the condenser to the water faucet and make sure that the hose from the top of the condenser is placed in the sink. Turn on the water so that it runs slowly.
   d. Carefully adjust the clamp holding the condenser so that you can connect the condenser *via* a cork stopper to the sidearm of the flask without placing strain on the sidearm.
   e. Connect the adapter to the lower end of the condenser.
   f. Wrap a piece of aluminum foil (not shown in Figure 3.2) around the visible part of the distilling flask and sidearm to provide insulation.
3. Place a 50 mL beaker under the adapter attached to the condenser. This beaker should be surrounded by ice as shown in Figure 3.2. Gently heat the liquid in the distilling flask with the heating mantle. Adjust the powerstat so that boiling takes place *slowly* and regularly. Vapors will rise up the neck of the flask, into the sidearm, and into the condenser, where they will be condensed to a liquid.

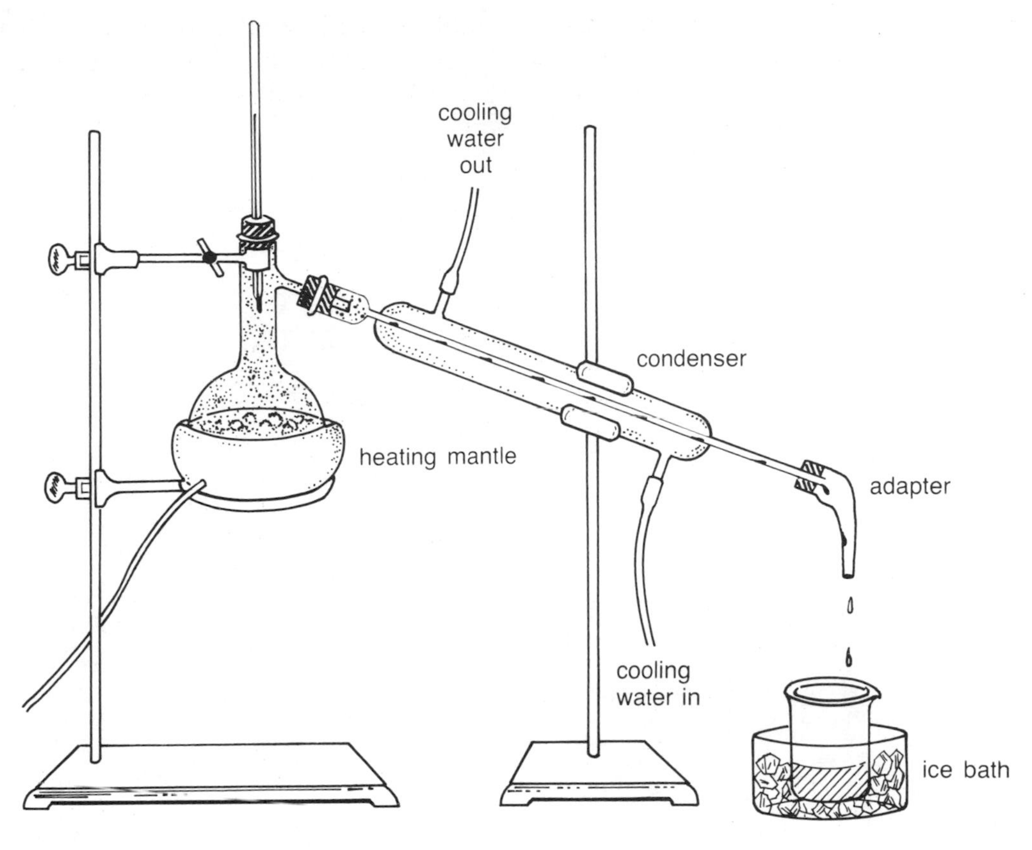

Figure 3.2

Record the temperature at which distillation takes place. Collect the liquid in the beaker until the temperature rises to a new boiling point and stabilizes. When this occurs, place another 50 mL beaker under the condenser to collect the higher-boiling component and record the new boiling point. Stop heating just before the distilling flask becomes empty. Use a 25 mL graduated cylinder to measure the volumes of the two fractions and record on the data sheet.

C. Extraction and Flotation

1. Test samples of sand, powdered sulfur, and sodium chloride for their behavior in water. Use your findings to devise a scheme for separating a mixture of these three materials. Describe your scheme on the data sheet.
2. Use a top-loader or open-beam balance to measure approximately 25 g of a mixture of sand, sulfur, and sodium chloride. Separate these from each other using the scheme you have devised. You may use beakers, funnel, filter paper, etc., as needed. The separated materials need not be dry. Record your observations during the separation. Show the separated samples to your laboratory instructor.

DATA SHEET

Name: ______________________ Lab Section: ______ Date: ____________

SEPARATION OF COMPONENTS OF A MIXTURE

A. Paper Chromatography

Distance from original ink spot to leading edge of:			**Ratio: (cm to color) / (cm to water)**
water		________ cm	
color	________	________ cm	________
color	________	________ cm	________
color	________	________ cm	________
color	________	________ cm	________

B. Distillation (Unknown number:______)

	Boiling Point	**Volume**
Low-boiling component	________	________
High-boiling component	________	________
	Total volume	________

Your mixture contains two of the compounds in the list below. They may be identified by their boiling points. The components of the mixture are:

______________________ and ______________________

Compound	**B.P.**	**Compound**	**B.P.**
acetone	56° C	*n*-propyl alcohol	97° C
methyl alcohol	65° C	*n*-butyl alcohol	118° C
ethyl alcohol	78° C		

C. Extraction and Flotation

1. Describe your scheme for separating the components of a mixture of sand, sulfur, and sodium chloride.

2. Observations made during separation:

QUESTIONS

1. Water is often purified by distillation. Describe how distillation of water can remove nonvolatile impurities such as salt, particles of leaves, and dirt.

2. Explain how washing clothes is analogous to the process of extraction.

3. Two solid substances are insoluble in water, and both have densities less than that of water. Could a mixture of these substances be separated by extraction or flotation using water? Explain your answer.

4. Chromatography is often used in forensic (crime detection) laboratories. What are some advantages of chromatography, compared to distillation or extraction, for isolating and identifying substances such as illegal drugs and poisons?

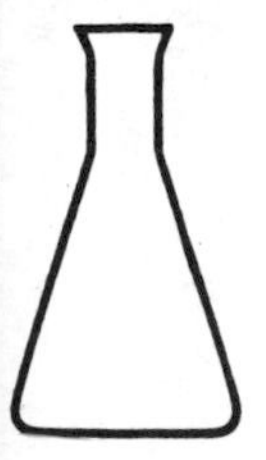

Chemical Reactions

OBJECTIVES

1. To learn to observe and record events in the laboratory and to draw conclusions from these observations.
2. To gain practice in writing and balancing chemical equations.

PRELABORATORY QUESTIONS

1. Explain what is meant by a "balanced" chemical equation.
2. When you mix two chemicals, how can you determine whether a chemical reaction has taken place?
3. What is a precipitate?
4. What is an indicator?

DISCUSSION

We live in a world where chemical reactions are continually taking place around and within us. The existence of life itself is dependent on numerous chemical reactions, such as photosynthesis (the production of carbohydrates by plants) and respiration (the burning of carbohydrates by animals to produce energy). We make use of chemical reactions to manufacture consumables, such as synthetic fibers, drugs, cosmetics, plastics, certain foods, food additives, detergents, fertilizers, and numerous others. Many of these reactions are very complex. Some familiar examples of simple reactions are combustion (burning) of gasoline, rusting of iron, and tarnishing of silver.

A chemical reaction is a process during which a *chemical change* occurs. A chemical change is one in which the *composition of matter* is altered. During a chemical reaction, the starting materials (reactants) combine to form products that differ in chemical composition from the reactants. Also, the physical properties of the products are often quite different from those of the reactants.

Chemical reactions may be described by chemical *equations* in which the reactants are written on the left, and the products are shown on the right. An arrow indicates the direction of the reaction. As an example, consider the reaction between the elements sodium (Na) and chlorine (Cl_2). These elements are the reactants, and the product of this reaction is sodium chloride (NaCl). The equation may be written:

$$Na + Cl_2 \longrightarrow NaCl \qquad \text{(not balanced)}$$

This equation is not *balanced* because there are two atoms of chlorine on the left side and only one on the right. The law of conservation of matter states that matter cannot be created or destroyed in a chemical reaction. Thus chemical equations must be balanced; that is, the number of atoms of each element on the left and right sides of the arrow must be the same.

When balancing equations, it is important not to change the *formulas* of the reactants and products. Instead, change the number of formula *units* used in the equation. In the above equation, 2 formula units of NaCl would provide the two chlorine atoms needed in the products:

$$Na + Cl_2 \longrightarrow 2\ NaCl \qquad \text{(not balanced)}$$

However, now there are two atoms of *sodium* on the right, so there should also be two on the left. The balanced equation is:

$$2\ Na + Cl_2 \longrightarrow 2\ NaCl$$

Chemical equations sometimes include designations such as (s) for solid, (l) for liquid, (g) for gas, and (aq) for aqueous solution. These are written directly after the symbol of an element or compound to indicate its physical state. For example, the above equation could be written:

$$2\ Na(s) + Cl_2(g) \longrightarrow 2\ NaCl(s)$$

However, in this experiment, where the emphasis is on learning how to write and balance equations, these symbols will be omitted.

Chemical changes are usually accompanied by observable physical changes such as a change in color, the formation of a *precipitate* (an insoluble substance), the evolution of a gas, or a change in temperature.

The above reaction is an example of a *synthesis* reaction. Another common type of reaction is *exchange*, where the cations and anions "exchange partners."

$$2\ NaBr + Pb(NO_3)_2 \longrightarrow 2\ NaNO_3 + \underline{PbBr_2}$$

Note that this equation is balanced by taking 2 formula units each of NaBr and $NaNO_3$. The line under $PbBr_2$ indicates that this is a precipitate. (Another way of indicating a precipitate is by an arrow: $PbBr_2 \downarrow$.)

How do we know that the precipitate we observe to form during the above reaction is $PbBr_2$ rather than $NaNO_3$? A simple test with a sample of $NaNO_3$ would show that it is quite *soluble* in water and therefore can be eliminated as a precipitate. The only other possible precipitate is $PbBr_2$. Since it would require considerable time for you to make all the tests necessary to identify the precipitates in this experiment, two general *solubility rules* are given here to assist you:

1. Compounds containing the cations Na^+, K^+, or NH_4^+ are soluble in water and therefore *cannot* be precipitates.
2. Compounds containing NO_3^- as the anion are soluble in water and *cannot* be precipitates.

During parts A and B of this experiment, you will mix pairs of chemical compounds (in aqueous solutions) together in small test tubes. In some cases a reaction will occur as indicated by a change in appearance of the mixture. In other cases there will be no evidence of reaction. You should make careful *observations* in each case and record your observations on the data sheet. In those cases where more than one compound reacts with the same reactant, try to identify the common component among the compounds.

In parts C and D you will test chemicals in solution with the *indicators* methyl orange and phenolphthalein. Indicators are chemical substances that change color under certain conditions. The indicators used in this experiment change color in the presence of *acids* (methyl orange) or *bases* (phenolphthalein). Carefully note any color changes during these experiments and try to deduce what chemical characteristics are common to acids and what characteristics are common to bases.

DATA SHEET

Name: ______________________ Lab Section: _____ Date: _________

CHEMICAL REACTIONS

MATERIALS NEEDED

small test tubes
phenolphthalein solution
methyl orange solution
0.1 M $AgNO_3$
0.1 M Na_2SO_4
0.1 M HCl
0.1 M NaCl
0.1 M $Ba(NO_3)_2$
0.1 M $BaCl_2$
saturated $Ba(OH)_2$
0.1 M HNO_3
0.1 M NaOH
0.1 M KOH
distilled water

EXPERIMENTAL PROCEDURE

A. Reaction with Silver Nitrate

CAUTION: Silver nitrate may cause dark spots on your skin upon contact. However, these spots are temporary.

In small, clean test tubes, add several drops of silver nitrate ($AgNO_3$) solution to five drops of the solutions listed below. Record your observations in the spaces provided.

If a reaction appears to take place (as indicated by formation of a cloudy precipitate), write a balanced chemical equation for the reaction. These will be *exchange* reactions, where cations and anions "exchange partners." Indicate precipitates by underlining (use the solubility rules given in the discussion). If no observable reaction occurs, write "N.R." (no reaction).

1. Hydrogen chloride, HCl (aqueous solution)

hydrogen chloride + silver nitrate

$$HCl + AgNO_3 \longrightarrow$$

2. Barium nitrate, $Ba(NO_3)_2$

barium nitrate + silver nitrate

$$Ba(NO_3)_2 + AgNO_3 \longrightarrow$$

3. Sodium hydroxide, NaOH

sodium hydroxide + silver nitrate

$$NaOH + AgNO_3 \longrightarrow$$

4. Sodium chloride, NaCl

sodium chloride + silver nitrate

$NaCl + AgNO_3 \longrightarrow$

5. Hydrogen nitrate, HNO_3 (aqueous solution)

hydrogen nitrate + silver nitrate

$HNO_3 + AgNO_3 \longrightarrow$

6. Barium hydroxide, $Ba(OH)_2$

barium hydroxide + silver nitrate

$Ba(OH)_2 + AgNO_3 \longrightarrow$

7. Barium chloride, $BaCl_2$

barium chloride + silver nitrate

$BaCl_2 + AgNO_3 \longrightarrow$

8. Potassium hydroxide, KOH

potassium hydroxide + silver nitrate

$KOH + AgNO_3 \longrightarrow$

The addition of $AgNO_3$ always results in a precipitate when certain ions are present in the solution. What are these ions?

B. Reaction with Sodium Sulfate

In small test tubes, add several drops of sodium sulfate (Na_2SO_4) solution to five drops of the solutions listed below. Record your observations in the spaces provided.

If a reaction appears to take place (as indicated by formation of a cloudy precipitate), write a balanced chemical equation for the reaction. These will be *exchange* reactions, where cations and anions "exchange partners." Indicate precipitates by underlining (use the solubility rules given in the discussion). If no observable reaction occurs, write "N.R." (no reaction).

1. Hydrogen chloride, HCl (aqueous solution)

hydrogen chloride + sodium sulfate

$HCl + Na_2SO_4 \longrightarrow$

2. Barium nitrate, $Ba(NO_3)_2$

barium nitrate + sodium sulfate

$Ba(NO_3)_2 + Na_2SO_4 \longrightarrow$

3. Sodium hydroxide, NaOH

sodium hydroxide + sodium sulfate

$NaOH + Na_2SO_4 \longrightarrow$

4. Sodium chloride, $NaCl$

sodium chloride + sodium sulfate

$NaCl + Na_2SO_4 \longrightarrow$

5. Hydrogen nitrate, HNO_3 (aqueous solution)

hydrogen nitrate + sodium sulfate

$HNO_3 + Na_2SO_4 \longrightarrow$

6. Barium hydroxide, $Ba(OH)_2$

barium hydroxide + sodium sulfate

$Ba(OH)_2 + Na_2SO_4 \longrightarrow$

7. Barium chloride, $BaCl_2$

barium chloride + sodium sulfate

$BaCl_2 + Na_2SO_4 \longrightarrow$

8. Potassium hydroxide, KOH

potassium hydroxide + sodium sulfate

$KOH + Na_2SO_4 \longrightarrow$

What ions are common to all the solutions that react with Na_2SO_4?

C. Reaction with Phenolphthalein Indicator

Put five drops of each of the following solutions into separate small test tubes half full of distilled water. Add *one drop* of phenolphthalein solution to each test tube and mix the solution. Record your observations. No equations are necessary.

1. Hydrogen chloride, HCl (aqueous solution)

2. Barium nitrate, $Ba(NO_3)_2$

3. Sodium hydroxide, $NaOH$

4. Sodium chloride, $NaCl$

5. Hydrogen nitrate, HNO_3 (aqueous solution)

6. Barium hydroxide, $Ba(OH)_2$

7. Barium chloride, $BaCl_2$

8. Potassium hydroxide, KOH

Phenolphthalein is an indicator for *bases*. Strong bases cause phenolphthalein to turn pink. List the solutions that gave a pink color with phenolphthalein:

What common ion do you find among the bases?

D. Reaction with Methyl Orange Indicator

Put five drops of each of the following solutions into separate small test tubes half full of distilled water. Add *one drop* of methyl orange solution to each test tube and mix the solution. Record your observations. No equations are necessary.

1. Hydrogen chloride, HCl (aqueous solution)

2. Barium nitrate, $Ba(NO_3)_2$

3. Sodium hydroxide, NaOH

4. Sodium chloride, NaCl

5. Hydrogen nitrate, HNO_3 (aqueous solution)

6. Barium hydroxide, $Ba(OH)_2$

7. Barium chloride, $BaCl_2$

8. Potassium hydroxide, KOH

Methyl orange is an indicator for *acids*. Strong acids cause methyl orange to turn red. List the solutions that appear to be acids:

What common ion do you find among the acids?

QUESTIONS

1. A precipitate is formed during each of the following reactions. Complete and balance the equations and indicate (by underlining) which of the possible products in each case is the precipitate.

$$Pb(NO_3)_2 + MgI_2 \longrightarrow$$
$$KOH + MgI_2 \longrightarrow$$
$$NH_4F + MgI_2 \longrightarrow$$

2. Which of the following compounds would cause phenolphthalein to turn pink? (Circle answers.)

K_2SO_4 $LiOH$ $Sr(OH)_2$ KBr
$MgCl_2$ NaI KCl $RbOH$ H_2SO_4

3. Which of the following compounds would cause methyl orange to turn red? (Circle answers.)

KNO_3 $LiOH$ H_2SO_4 $CaCl_2$
MgI_2 $HClO_4$ HBr $NaBr$

EXPERIMENT 5

Determination of the Formulas of Precipitates

OBJECTIVES

1. To provide an opportunity for students to apply simple logic in interpreting observations made in the laboratory.
2. To gain practice in writing and balancing chemical equations.

PRELABORATORY QUESTIONS

1. What is an exchange reaction? Give an example.
2. Explain why many of the boxes in the grid on the data sheet are crossed off.

DISCUSSION

Chemists frequently find that their job is similar to that of a detective because, like detectives, they are required to exercise their powers of *observation* and *logic* in order to interpret facts and solve problems. In this experiment you will use simple logic to determine the formulas of precipitates that occur when pairs of reagents are mixed in solution. Since the reactions are of the *exchange* type, where cations "trade" anions ($AB + CD \longrightarrow AD + CB$), any precipitate formed must be one of the two products, AD or CB, in the reaction. By comparing the results of a series of related reactions, you will have sufficient information to deduce which of the products in each pair is the precipitate.

MATERIALS NEEDED

glass square
marking pen

dropper bottles containing the following 0.1 M solutions:

Kit 1	Kit 2	Kit 3	Kit 4
$NaCl$	K_2CrO_4	$NaCl$	$FeCl_3$
Na_2SO_4	$AgNO_3$	$NaBr$	$CoCl_2$
$BaCl_2$	KCl	$AgNO_3$	$Co(NO_3)_2$
KNO_3	K_2SO_4	Na_2SO_4	$NaOH$
$AgNO_3$	$BaCl_2$	$BaCl_2$	KOH
KCl	KNO_3	$NaNO_3$	KCl

EXPERIMENTAL PROCEDURE

Your instructor will make available four kits, each of which contains six solutions in dropper bottles. Use one kit at a time and make sure each bottle is returned to its proper kit. There is a separate data sheet for each kit.

In the spaces labeled A to F along the top and on the left side of the grid on one of the data sheets, write the formulas of reagents (in any order) from one of the kits. Note that boxes on the grid that correspond to the reaction of any compound with *itself* (e.g., A + A) are crossed off, as are also those squares that correspond to the *same* reaction (e.g., A + B, B + A).

Mark off a 3 × 3 grid (tic-tac-toe) on a glass square with a marking pen. Turn the glass over and place one drop of reagent A in as many spaces on the glass as there are open boxes on the data sheet under A. Add one drop of reagent B to one of these drops, one drop of reagent C to another, etc., until all the reagents have been mixed with reagent A. Observe what happens in each case. If a precipitate forms, write PPT in the proper box on the data sheet and indicate the color. If no precipitate forms, write NO. Wash and dry the glass square and repeat the procedure, beginning with reagent B. Continue in this manner until all combinations of reagents in the kit have been mixed.

When you have finished with the kit, examine your data to see what combinations of ions in the solutions mixed *could* give the observed precipitates (e.g., if Na_2SO_4 + $BaCl_2$ gives a precipitate, the possible precipitates are $NaCl$ and $BaSO_4$). Make a list of these pairs of possible precipitates in the space provided near the bottom of the data sheet.

The experiment is designed so that whenever precipitation occurs, only *one* of the two possible products is a precipitate. Examine the formulas of the *solutions* in the kit to see if any of these are listed among the possible precipitates; if so, cross these from your list. This should reveal the formula of the precipitate in at least one of the pairs of possible precipitates. If this same formula occurs elsewhere in your list, it will also be the precipitate in that reaction, and the other member of the pair must be a soluble compound.

After carefully examining your data and applying logic in this manner, you should be able to deduce which compound is the precipitate in each pair of possibilities in your list. Circle the precipitates in your list and write balanced equations for the reactions that give precipitates (underline the precipitate in each equation).

When you have finished with one kit, choose another and repeat the above procedure. You should complete all four kits during the laboratory period.

DATA SHEET

Name: ______________________ Lab Section: _____ Date: ___________

DETERMINATION OF THE FORMULAS OF PRECIPITATES

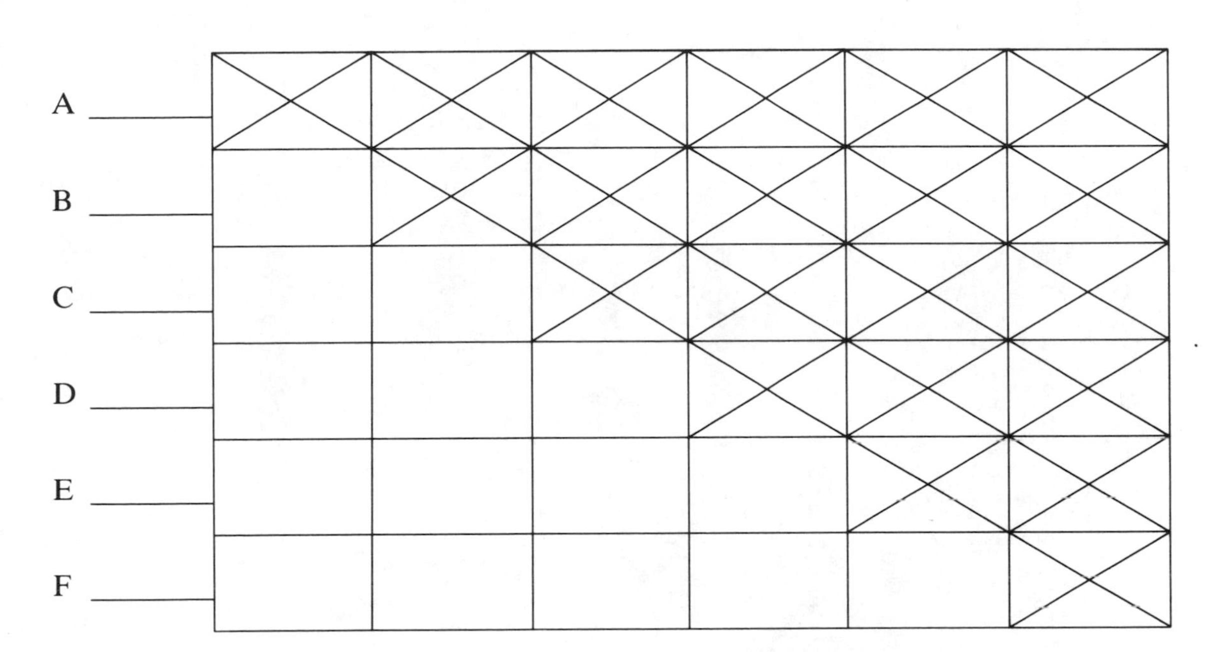

Possible precipitates:

Equations for precipitation:

DATA SHEET

Name: ______________________ Lab Section: _____ Date: __________

DETERMINATION OF THE FORMULAS OF PRECIPITATES

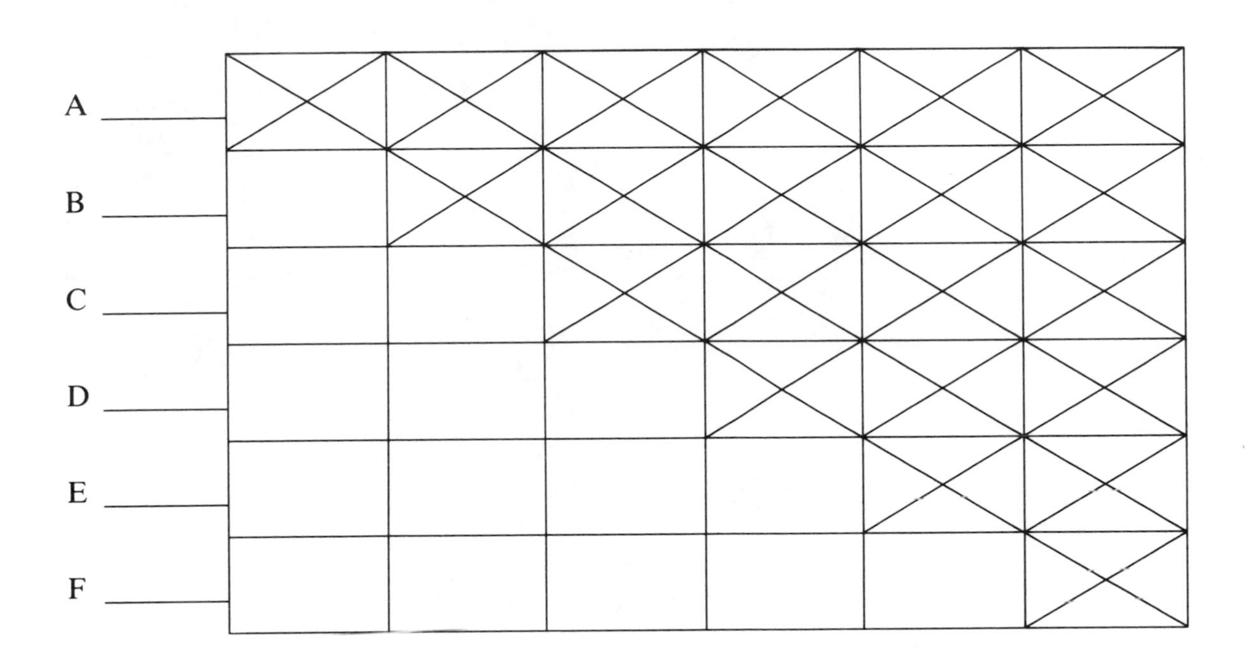

Possible precipitates:

Equations for precipitation:

DATA SHEET

Name: ______________________________ Lab Section: _____ Date: __________

DETERMINATION OF THE FORMULAS OF PRECIPITATES

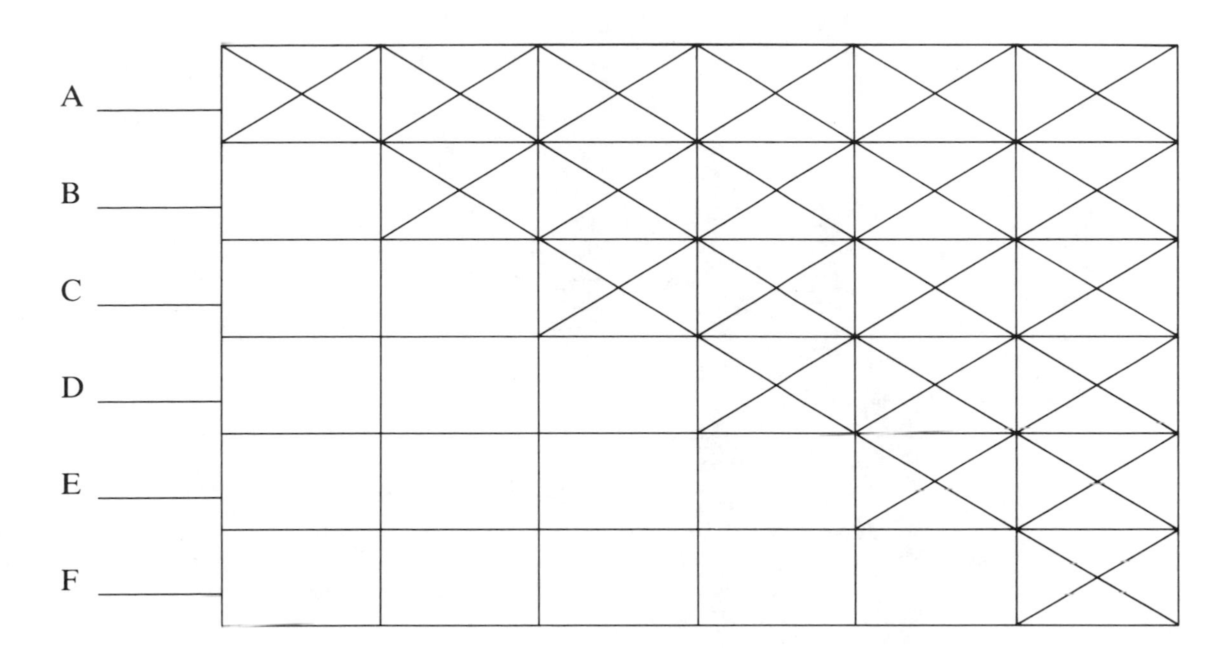

Possible precipitates:

Equations for precipitation:

DATA SHEET

Name: ______________________ Lab Section: ______ Date: __________

DETERMINATION OF THE FORMULAS OF PRECIPITATES

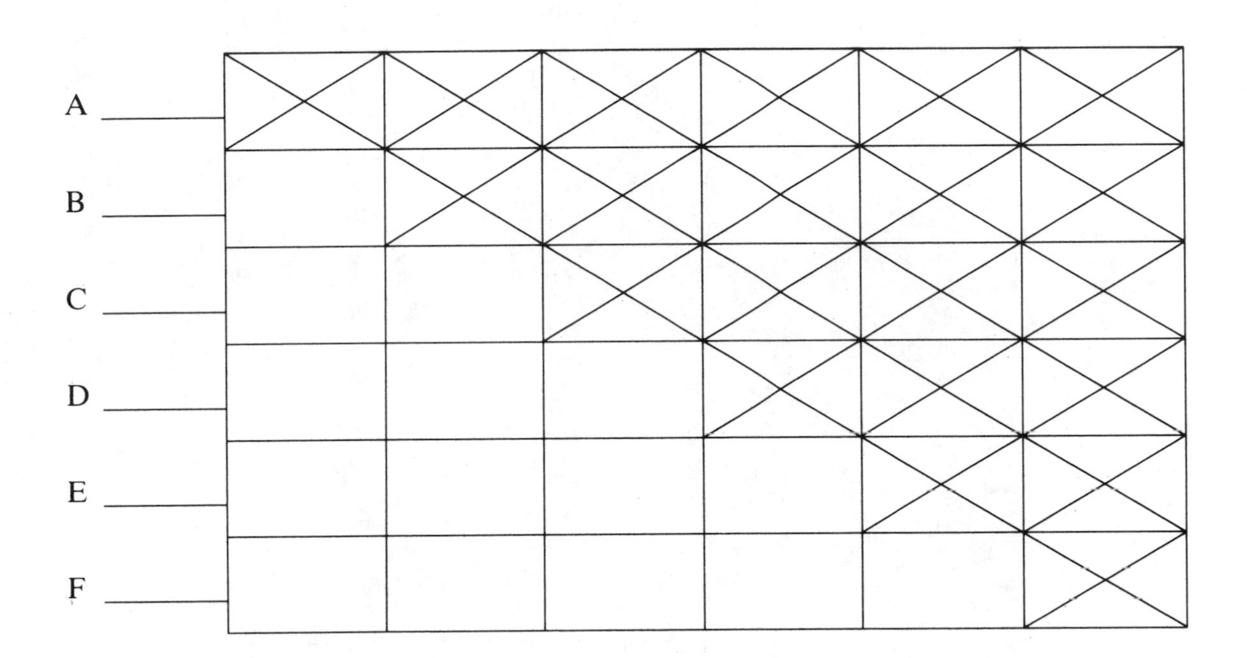

Possible precipitates:

Equations for precipitation:

QUESTIONS

1. A precipitate is observed for each of the following reactions. Use the information on your data sheets to determine the formula of the precipitate in each case, then complete and balance the equations. Indicate the precipitate in each equation by underlining.

$MgCl_2 + NaOH \longrightarrow$

$Pb(NO_3)_2 + KBr \longrightarrow$

$NaF + Ba(NO_3)_2 \longrightarrow$

$MgSO_4 + NaOH \longrightarrow$

2. It is known that compounds of certain cations are nearly always soluble in water, regardless of the anions present. Similarly, compounds of certain anions are nearly always soluble. For which cations and anions did you find this to be true during this experiment?

EXPERIMENT 6

Determination of the Equation of a Chemical Reaction

OBJECTIVES

1. To demonstrate how the coefficients in a balanced chemical equation may be determined experimentally.
2. To illustrate the use of a catalyst in a chemical reaction.

PRELABORATORY QUESTIONS

1. What is a catalyst?
2. Explain why you should not touch the crucible or cover with your fingers during the experiment.
3. Why is there a loss in mass during the decomposition of potassium chlorate?

DISCUSSION

In a balanced chemical equation, the coefficients show the *mole ratios* of reactants and products. These coefficients can be determined experimentally if the number of *grams* of reactants consumed and the number of *grams* of products formed during a reaction can be measured, because these masses can be converted to *moles*.

The thermal decomposition of potassium chlorate ($KClO_3$) gives oxygen gas (O_2) and another product. In this experiment, you will determine both the identity of the other product and the coefficients of the balanced chemical equation for the reaction. The decomposition of $KClO_3$ proceeds more rapidly and at a lower temperature if a *catalyst*, manganese dioxide (MnO_2), is mixed with the $KClO_3$. A catalyst is a substance that increases the rate of a reaction but is not changed by the reaction.

A crucible containing MnO_2 and $KClO_3$ is weighed accurately. Then the mixture is heated to carry out the reaction, after which the crucible is cooled and reweighed. The loss in weight reveals the mass of O_2 formed during the reaction, and this can be converted to *moles* of O_2.

It may be assumed that *all* the $KClO_3$ decomposes, so that the initial mass of $KClO_3$ represents the mass of $KClO_3$ consumed by the reaction. This may be converted into *moles* of $KClO_3$. From these quantities, the *ratio* of moles of O_2 formed to moles of $KClO_3$ consumed in the reaction may be determined. This ratio helps identify the other product of the reaction and ultimately leads to the complete, balanced equation.

MATERIALS NEEDED

- ringstand with ring
- gas burner
- clay pipestem triangle
- crucible with cover
- crucible tongs
- watch glass
- small test tubes
- small beaker
- stirring rod
- filter funnel
- filter paper
- MnO_2 (reagent grade)
- $KClO_3$ (reagent grade)
- 1 M HNO_3
- 0.1 M KCl
- 0.1 M $AgNO_3$

EXPERIMENTAL PROCEDURE

A. Decomposition of Potassium Chlorate

CAUTION: $KClO_3$ is a strong oxidizing agent. It is capable of causing *fires* or *explosions* when mixed with acids, ammonium salts, or substances that are easily oxidized (combustible substances). Therefore, follow these precautions during the experiment:

- Wear safety goggles at all times.
- Use only *reagent grade* $KClO_3$ and MnO_2.
- Make sure that your crucible is *clean* before you begin.
- If any $KClO_3$ is spilled, *do not* discard it in a trash can. Instead, wipe it up with a *wet* towel and wash the towel *thoroughly* at the sink.

1. Place a small amount (between 0.2 and 0.5 g) of reagent grade manganese dioxide (MnO_2) in a *clean*, dry crucible.
2. Attach a ring to a ringstand so that the ring is about 5 cm above the top of a gas burner placed on the base of the ringstand. Lay a clay pipestem triangle on the ring, then put the crucible on the triangle and cover it (see Figure 6.1). Light the burner and adjust it so that the top of the blue portion of the flame is about 1 cm below the bottom of the crucible. Heat the crucible to red heat for about 5 minutes. This is to drive off moisture from the crucible and MnO_2 prior to accurate weighing.

   From this point until the completion of step 6, handle the crucible and

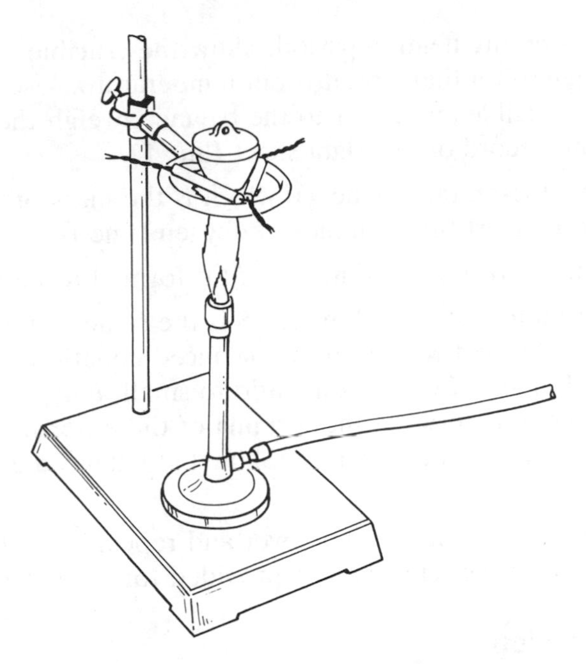

Figure 6.1

cover only with crucible tongs. The crucible and cover should be placed *only* on the clay triangle, the watch glass, or the balance pan. These procedures are necessary to prevent the crucible from gaining weight by picking up material from the bench top or moisture and oil from your fingers.

3. Allow the crucible and cover to cool for at least 10 minutes. If you feel any warmth when the back of your hand is brought near, allow more cooling time. *Do not touch the crucible or cover with your fingers*. When the crucible and cover are cool, use the *tongs* to place them on the watch glass, then carry them to an analytical balance. Weigh the crucible, cover, and MnO_2 to the nearest 0.0001 g and record on the data sheet (line b).

4. Use a top-loader or open-beam balance to weigh between 1.5 and 2.0 g of reagent grade potassium chlorate ($KClO_3$) into a clean, small beaker. Transfer the $KClO_3$ into the crucible and mix the $KClO_3$ and MnO_2 as best you can by rotating the crucible carefully while holding it with tongs. Reweigh the crucible, cover, and contents to the nearest 0.0001 g and record on the data sheet (line a).

5. Place the crucible and cover on the clay triangle and light the burner. Hold the base of the burner in your hand and move the burner back and forth so that the flame *gently* heats the crucible. Heating too strongly at this point may cause some of the reactants to be expelled from the crucible. Continue this until the reaction mixture changes from a bubbling liquid to a gray solid (*carefully* remove the crucible cover with tongs to observe this). Then place the burner below the crucible and heat to red heat for at least 25 minutes.

   While waiting for the reaction to finish, calculate the mass (line c) and number of moles (line e) of $KClO_3$ present at the beginning of the reaction. Part B(1,2) of the experiment may also be carried out during this period.

6. After the 25-minute heating period, allow the crucible and cover to cool on the triangle until they are at room temperature. Use the watch glass to carry the crucible and cover to the balance, weigh them to the nearest 0.0001 g, and record on the data sheet (line f).
7. Calculate the loss in mass (line g), which is the mass of oxygen (O_2) formed, and convert this to moles of oxygen (line i).
8. Calculate the *ratio* (line j) of moles of O_2 formed to moles of $KClO_3$ used.
9. The ratio of moles calculated in step 8 is the same as the ratio of the coefficients of O_2 and $KClO_3$ in the balanced equation (allowing for some experimental error). Convert this ratio to small, *integral* coefficients. You should now be able to write the formula of the other product of the reaction and balance the equation. Show the balanced equation on the data sheet.
10. Wash and dry your crucible and cover and repeat steps 1–9 above. A second column in the data sheet is provided for your data.

B. Tests for Chloride Ion

1. To a small test tube containing approximately 1 mL of 0.1 M KCl, add several drops of 1 M HNO_3 and five drops of 0.1 M $AgNO_3$.

   CAUTION: $AgNO_3$ solution may cause dark spots (which are only temporary) on your skin upon contact.

   Describe your observations on the data sheet and write a balanced equation for the reaction. (The HNO_3 is used to acidify the solution and is *not* needed in the equation.)
2. Place several crystals of $KClO_3$ in a small test tube and add approximately 1 mL of distilled water to dissolve. Add several drops of 1 M HNO_3 and five drops of 0.1 M $AgNO_3$. Record your observations on the data sheet.
3. After the crucible and cover have been weighed in part A, step 6, add several milliliters of distilled water to the crucible and stir the mixture with a stirring rod. Pour the mixture onto a piece of filter paper, which has been folded and placed in a funnel, and collect the *filtrate* (the liquid that passes through the filter) in a small beaker. Describe the material remaining on the filter paper. What is this material?
4. Add several drops of 1 M HNO_3 and five drops of $AgNO_3$ to approximately 1 mL of the filtrate. Record your observations on the data sheet.

DATA SHEET

Name: ______________________ Lab Section: ______ Date: ____________

DETERMINATION OF THE EQUATION OF A CHEMICAL REACTION

A. Decomposition of Potassium Chlorate

*a. Mass of crucible, cover, MnO_2, and $KClO_3$ before reaction	______	______
*b. Mass of crucible, cover, and MnO_2 before reaction	______	______
c. Mass of $KClO_3$ (line a − line b)	______	______
d. Formula mass of $KClO_3$ (grams/mole)	122.5	122.5
e. Moles of $KClO_3$ (line c/line d)	______	______
*f. Mass of crucible, cover, and contents after reaction	______	______
g. Mass of O_2 formed (line a − line f)	______	______
h. Molecular mass of O_2 (grams/mole)	32.00	32.00
i. Moles of O_2 evolved (line g/line h)	______	______
j. Ratio: (moles O_2)/(moles $KClO_3$)	______	______

Balanced equation for thermal decomposition of $KClO_3$:

B. Tests for Chloride Ion

1. Observations:

Balanced equation (this is an exchange reaction):

* Experimental data

2. Observations:

3. Describe the material remaining on the filter paper:

What is this material?

4. Describe what happens when $AgNO_3$ solution is added to the filtrate from the $KClO_3$ reaction:

Using results from the above test, propose a formula for the other product of the $KClO_3$ reaction:

Does this agree with your balanced equation?

QUESTIONS

1. a. Balance the following equation:

$Na + O_2 \longrightarrow Na_2O$

b. What is the mole ratio of Na to O_2?

c. How many moles of O_2 are needed to react with 2 moles of Na?

d. How many *grams* of O_2 are needed to react with 2 moles of Na?

e. How many moles of Na_2O could be produced from 2 moles of Na?

f. How many *grams* of Na_2O could be formed from 2 moles of Na?

2. During the reaction of P with O_2, the combining mole ratio of P to O_2 is 1.0 to 1.25. Assuming only one product is formed, write a complete, balanced equation for this reaction.

Acids, Bases, Buffers, and pH

OBJECTIVES

1. To gain an understanding of the relationship between pH and concentration of acids and bases.
2. To observe the differences in properties of strong acids and weak acids.
3. To learn how a buffer is prepared and to observe how it functions.

PRELABORATORY QUESTIONS

1. Describe the difference between a strong acid and a weak acid. Give an example of each.
2. What does pH measure? What are the normal pH ranges for acidic and basic solutions?
3. Explain what is meant by the term *buffer*.

DISCUSSION

Acids are defined as substances capable of *donating protons*, whereas bases are substances capable of *accepting protons*.

In our daily activities we encounter many acidic or basic substances, and acids and bases play important roles in our metabolism. Many common foods are acidic. Examples are citrus fruits, jams, jellies, coffee, cider, soft drinks, cabbage, sauerkraut, vinegar, and milk. A few foods, such as egg whites, are basic. Gastric juice is very acidic, and we occasionally take an antacid (a base) to relieve indigestion. Milk of magnesia, a laxative, is basic, as are most cleaning agents, including soaps, ammonia, and lye. Since high concentrations of acids or bases are harmful to living organisms, nature has developed mechanisms for precise control (buffering) of the concentrations of acids and bases in body fluids such as blood.

A *strong acid* is one that dissociates (breaks up) completely into ions when dissolved in water. Examples of strong acids are HCl, HNO_3, and H_2SO_4. These compounds react with water to form the *hydronium ion*, H_3O^+, and an anion characteristic of the acid. For HCl:

$$HCl + H_2O \longrightarrow H_3O^+ + Cl^-$$

This reaction goes completely to the right, i.e., a mole of HCl produces a mole of H_3O^+ and a mole of Cl^-.

Similarly, a *strong base* dissociates completely in water to form the *hydroxide ion*, OH^-, and a cation characteristic of the base. NaOH and KOH are strong bases. In water:

$$NaOH \longrightarrow Na^+ + OH^-$$

A mole of NaOH produces a mole of Na^+ ions and a mole of OH^- ions.

The *pH* of an acidic or basic solution is a measure of the concentration of hydronium ion present. pH is defined as follows:

$$pH = -\log [H_3O^+] \qquad (1)$$

Since this is a logarithmic expression, a change in *one* pH unit corresponds to a *tenfold* change in $[H_3O^+]$. Acids normally have pH values between 0 and 7, whereas bases have pH values between 7 and 14. A solution having a pH of 7 is said to be "neutral."

For a strong acid, $[H_3O^+]$ is the same as the concentration of the acid itself. As an example, a 0.01 M solution of HCl has $[H_3O^+] = 0.01$ or 1.0×10^{-2}, and the pH is 2. Similarly, a 0.001 M solution of HCl has a pH of 3.

For a strong base, the concentration of OH^- is the same as that of the base. The product of $[OH^-]$ and $[H_3O^+]$ is always constant (the "ion product of water"):

$$[H_3O^+]\,[OH^-] = 1.0 \times 10^{-14} \qquad (2)$$

Thus, $[H_3O^+]$ may be determined by substituting $[OH^-]$ in equation 2. Once $[H_3O^+]$ is found, the pH of the basic solution may be determined using equation 1. For example, a 0.01 M solution of NaOH has an $[OH^-]$ of 1.0×10^{-2}. Using equation 2, $[H_3O^+]$ is found to be 1.0×10^{-12}. Thus, the pH of this basic solution is 12 (from equation 1).

Weak acids and bases dissociate only to a small extent in water. Acetic acid, CH_3COOH, the active component of vinegar, is a typical weak acid. In water:

$$CH_3COOH + H_2O \rightleftharpoons H_3O^+ + CH_3COO^-$$

The double arrow in this equation shows that the reaction goes in both directions. The larger arrow pointed toward the left indicates that the reaction to the left goes much faster. As a result, there are relatively few H_3O^+ and CH_3COO^- ions present in solution, and unlike a strong acid, a 0.01 M solution of acetic acid does *not* have a pH of 2.0. Instead, the pH is higher, indicating that fewer H_3O^+ ions are present.

The ratio of moles of ions present to moles of the neutral acid, CH_3COOH, is given by the *equilibrium constant expression* for the acid. For acetic acid this is:

$$K_a = \frac{[H_3O^+][CH_3COO^-]}{[CH_3COOH]} = 1.8 \times 10^{-5} \tag{3}$$

This expression may be used in calculating the pH of a solution of acetic acid. Although very little CH_3COOH dissociates, every mole of CH_3COOH that does dissociate produces a mole of H_3O^+ *and* a mole of CH_3COO^-. Thus, $[H_3O^+] = [CH_3COO^-]$, and the above equation may be rearranged:

$$[H_3O^+]^2 = [CH_3COOH] \times (1.8 \times 10^{-5}) \tag{4}$$

If $[CH_3COOH]$ is known, we can solve for $[H_3O^+]$, then use equation 1 to solve for the pH.

Buffer solutions are solutions that are resistant to *changes* in pH. A typical buffer is a solution of a weak acid, such as CH_3COOH, *and* a salt of the weak acid, such as $Na^+CH_3COO^-$. If a small amount of *base* is added to this buffer solution, it reacts with the weak acid present:

$$NaOH + CH_3COOH \longrightarrow Na^+CH_3COO^- + H_2O$$

The added base is neutralized; thus the pH of the buffer solution does not change significantly. If a small amount of *acid* is added to the buffer solution, the acid will be consumed by $Na^+CH_3COO^-$ (which acts as a base):

$$Na^+CH_3COO^- + HCl \longrightarrow CH_3COOH + NaCl$$

Here, some CH_3COOH is formed, but the pH does not change significantly because CH_3COOH dissociates only to a small extent. Of course, if the amount of base or acid added to the buffer is sufficiently *large* to consume most of the CH_3COOH or $Na^+CH_3COO^-$ present, the pH will change. In this case the *capacity* of the buffer has been exceeded.

During this experiment, you will measure the pH's of some materials commonly found in the household, and you will observe relationships between concentration and pH for a strong acid, a strong base, and a weak acid. Also, you will determine the effects of added acid and base on the pH of a buffer solution.

DATA SHEET

Name: ______________________ Lab Section: ______ Date: __________

ACIDS, BASES, BUFFERS, AND pH

MATERIALS NEEDED

small test tubes	0.1 M HCl
10 mL graduated cylinder	0.1 M NaOH
100 mL graduated cylinder	0.001 M NaOH
pHydrion indicator solution	0.1 M acetic acid
solutions of household items	0.2 M acetic acid
distilled water	0.2 M sodium acetate

EXPERIMENTAL PROCEDURE

A. pH Values of Some Common Household Materials

Solutions of several items commonly found in the household will be provided by your laboratory instructor. To four mL of each of these solutions in small test tubes, add one drop of indicator solution and *mix well*. Determine the pH of the solution by comparing its color with the color chart provided with the indicator. Record the pH values in the spaces provided and indicate by the letters A, B, and N whether these solutions are acidic, basic, or neutral, respectively.

1. Ammonia pH = __________
2. Lemon juice pH = __________
3. Baking soda pH = __________
4. Dishwasher detergent pH = __________
5. Table salt pH = __________
6. Oven cleaner pH = __________
7. Vinegar pH = __________
8. Household cleaner pH = __________

B. pH Values of Acids, Bases, and Buffers

1. *HCl, A Strong Acid*

a. Add one drop of indicator solution to 4 mL of 0.1 M HCl in a small test tube and *mix well*. Determine the pH of the HCl solution by comparing its color with the color chart provided with the indicator. Keep this solution for comparison with those in parts b and c, below. Use the concentration of HCl to determine $[H_3O^+]$. What pH is *expected*, based on the concentration of the acid?

observed pH = ________

[HCl] = 0.1 M

$[H_3O^+]$ = ________

expected pH = ________

b. Rinse a 10 mL graduated cylinder thoroughly with distilled water and use the cylinder to dilute *1* mL of 0.1 M HCl to 10 mL, using distilled water. To 4 mL of the diluted solution add one drop of indicator and *mix well*. Record the pH. *Save* the remainder of the diluted solution for part c, below. Calculate and record the concentrations of HCl and H_3O^+. What pH is expected, based on the concentration of the acid?

observed pH = ________

[HCl] = ________

$[H_3O^+]$ = ________

expected pH = ________

c. Rinse the 10 mL graduated cylinder again with distilled water, and dilute *1* mL of the remaining solution from part b to 10 mL, using distilled water. To 4 mL of the diluted solution add one drop of indicator and *mix well*. Record the pH. Calculate and record the concentrations of HCl and H_3O^+. What pH is expected, based on the concentration of the acid?

observed pH = ________

[HCl] = ________

$[H_3O^+]$ = ________

expected pH = ________

d. Are the observed pH values consistent with what you would expect from the H_3O^+ concentrations? Explain your answer.

e. How do the pH values change with each tenfold dilution of the acid? Is this change expected?

2. *NaOH, A Strong Base*

a. Add one drop of indicator solution to 4 mL of 0.001 M NaOH and *mix well.* Determine the pH of the NaOH solution by comparing its color with the color chart provided with the indicator. Use the NaOH concentration to determine $[OH^-]$. Use equation 2 to determine $[H_3O^+]$, then calculate the expected pH.

observed pH = ________

[NaOH] = 0.001

$[OH^-]$ = ________

$[H_3O^+]$ = ________

expected pH = ________

b. Is the observed pH consistent with what you expect based on the H_3O^+ concentration? Explain.

c. If the solution in part a were diluted tenfold, what should be the pH?

d. If the solution in part a were concentrated tenfold, what should be the pH?

3. *CH_3COOH, A Weak Acid*

a. Add one drop of indicator solution to 4 mL of 0.1 M CH_3COOH and *mix well.* Determine the pH of the CH_3COOH solution by comparing its color with the color chart provided with the indicator. Keep this solution for comparison with part d, below. Calculate $[H_3O^+]$ for this solution by using equation 4 (round off the answer to one significant digit), then calculate the expected pH.

observed pH = ________

$[CH_3COOH]$ = 0.1

$[H_3O^+]$ = ________

expected pH = ________

b. Is the observed pH consistent with what you would expect from the H_3O^+ concentration? Explain.

c. What would be the pH of this solution if it were a *strong* acid (0.1 M)?

d. Rinse a *100 mL* graduated cylinder thoroughly with distilled water and use the cylinder to dilute *1* mL of 0.1 M CH_3COOH to 100 mL using distilled water. To 4 mL of the diluted solution, add one drop of indicator and *mix well.* Determine the pH. Record the concentration of CH_3COOH. Calculate $[H_3O^+]$ from equation 4 (round off to one digit), then calculate the expected pH.

observed pH = ________

$[CH_3COOH]$ = ________

$[H_3O^+]$ = ________

expected pH = ________

e. Is the observed pH consistent with what you would expect from the H_3O^+ concentration? Explain.

f. What would be the pH of this solution if it were a *strong* acid having the above molar concentration?

4. *CH_3COOH and $Na^+CH_3COO^-$, A Buffer Solution*

a. Mix 10 mL of 0.2 M CH_3COOH with 10 mL of 0.2 M $Na^+CH_3COO^-$. This makes 20 mL of a buffer solution containing 0.1 M CH_3COOH and 0.1 M $Na^+CH_3COO^-$.

b. Add one drop of indicator solution to 4 mL of the buffer solution and *mix well*. Determine the pH of the buffer solution by comparing its color with the color chart provided with the indicator. Keep this solution to compare with those in parts c–f.

observed pH = ________

c. Add one drop of 0.1 M HCl to 4 mL of the buffer solution. Then add one drop of indicator to this solution, mix, and determine the pH.

observed pH = ________

d. Add another nine drops of 0.1 M HCl (total of ten drops) to the solution in *part c*, mix, and determine the pH.

observed pH = ________

e. Add one drop of 0.1 M NaOH to 4 mL of the buffer solution. Then add one drop of indicator to this solution, mix, and determine the pH.

observed pH = ________

f. Add another nine drops of 0.1 M NaOH (total of ten drops) to the solution in *part e*, mix, and determine the pH.

observed pH = ________

g. What is the change, if any, in the pH of the buffer solution upon addition of:

1 drop of acid? ________

10 drops of acid? ________

1 drop of base? ________

10 drops of base? ________

h. Does the solution of acetic acid and sodium acetate behave as a buffer? Explain.

i. Wash a test tube thoroughly and fill it with 4 mL of distilled water. Add one drop of indicator solution, mix, and determine the pH.

observed pH = ________

j. Add one drop of 0.1 M HCl to 4 mL of distilled water. Then add one drop of indicator to this solution, mix, and determine the pH.

observed pH = ________

k. Add another nine drops of 0.1 M HCl (total of ten drops) to the solution in *part j*, mix, and determine the pH.

observed pH = ________

l. Add one drop of 0.1 M NaOH to 4 mL of distilled water. Then add one drop of indicator to this solution, mix, and determine the pH.

observed pH = ________

m. Add another nine drops of 0.1 M NaOH (total of ten drops) to the solution in *part l*, mix, and determine the pH.

observed pH = ________

n. What is the *change*, if any, in pH of the unbuffered water upon addition of:

1 drop of acid? ________

10 drops of acid? ________

1 drop of base? ________

10 drops of base? ________

o. Does distilled water behave as a buffer? Explain.

p. Compare the buffer solution and the unbuffered water with regard to the effects of acid and base.

QUESTIONS

1. A solution of a certain strong acid has a pH of 3. One mL of this solution is diluted to 1000 mL with distilled water. What is the pH of the diluted solution?

2. Determine the pH values of the following solutions:

   a. 1.0×10^{-6} M HNO_3

   b. 1.0×10^{-6} M KOH

   c. 5.6×10^{-4} M CH_3COOH

3. *Para*-chlorobenzoic acid, ClC_6H_4COOH, is a weak acid with an equilibrium constant K_a of 1.0×10^{-4}.

   a. Write an equation (similar to that shown for acetic acid in the discussion) for the *dissociation* of *para*-chlorobenzoic acid.

b. Write the *equilibrium constant expression* for ClC_6H_4COOH (analogous to equation 3).

c. Derive an equation for ClC_6H_4COOH that is analogous to equation 4 (for acetic acid). Use your equation to determine the pH of a 1.0×10^{-2} M solution of ClC_6H_4COOH.

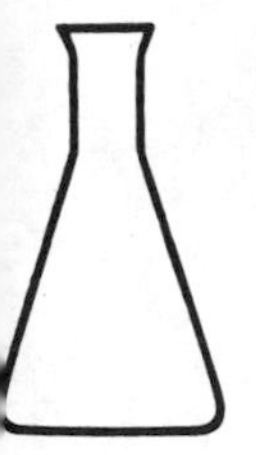

Titration of Vinegar

OBJECTIVES

1. To demonstrate a simple method of chemical analysis.
2. To provide experience with calculations involving molar quantities and solution concentrations.

PRELABORATORY QUESTIONS

1. What is the purpose of carrying out a titration?
2. Why is the indicator phenolphthalein used?
3. Describe how to calculate the number of moles of NaOH added to the vinegar solution.

DISCUSSION

Clinical laboratories daily perform hundreds of *analyses* to find out how much of a particular chemical compound is present in an "unknown" sample containing many other substances. The amount of compound in question is usually reported in units of milligrams per deciliter in clinical laboratories, but other units of concentration, such as parts per million, percent by weight, or moles per liter could also be used. Most of these analyses are carried out using electronic instruments and automated routines.

Chemists frequently use a method of analysis known as *titration*. This method requires relatively large samples but simple apparatus. In this technique, a solution of *known* concentration of a reagent that reacts with the compound in question is added in small increments to a solution containing a measured amount of the unknown sample. The reagent is added until all of the unknown compound has reacted. This "end point" is indicated by a *change* in some readily recognizable physical property, such as color, of either the unknown or the analytical reagent. Alternatively, a third compound, an *indicator*, whose color depends on which of the other substances is in excess, may be added.

In this experiment you will use a buret (shown in Figure 8.1) to add increments of the reagent to the unknown. You will titrate an acid (acetic acid of unknown concentration in white vinegar) with a base (a sodium hydroxide solution of known concentration). The acetic acid, CH_3COOH, reacts with sodium hydroxide to give water and sodium acetate:

$$CH_3COOH + NaOH \longrightarrow H_2O + Na^+CH_3COO^-$$

The number of *moles* of NaOH added to the vinegar sample can be determined by measuring the volume of NaOH solution delivered by the buret and by knowing its concentration:

$$n \text{ (moles)} = \text{volume (L)} \times \text{concentration (moles/L)} \qquad (1)$$

Note that the volume must be expressed in *liters* in this equation.

Since each mole of NaOH added reacts with exactly 1 mole of CH_3COOH, n will also equal the number of moles of CH_3COOH present in the unknown solution. From the number of moles of CH_3COOH and its formula mass (60.05 g/mole), the number of *grams* of CH_3COOH present in the unknown can be calculated:

$$\text{g} = \text{formula mass (g/mole)} \times n \text{ (moles)} \qquad (2)$$

Finally, the mass of CH_3COOH present may be divided by the total mass of the unknown sample to determine the *percent by weight* of acetic acid present in the sample:

$$(\text{g } CH_3COOH)/(\text{g vinegar}) \times 100 = \%\ CH_3COOH \qquad (3)$$

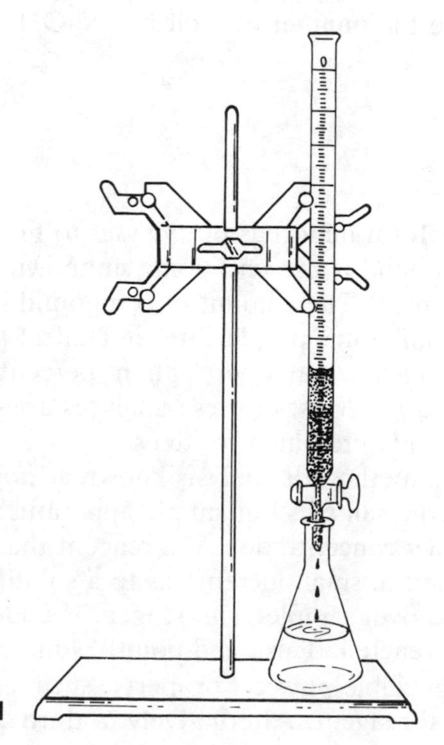

Figure 8.1

Since both reagents in this titration are colorless, it is necessary to add an indicator, *phenolphthalein*, which is colorless in acid solution but pink in basic solution. The indicator will be added to the vinegar solution, and sodium hydroxide will be added from the buret until the solution shows the faintest detectable pink color. This indicates that sodium hydroxide is in slight excess and that the end point of the titration has been reached. Several titrations will be performed, using fresh samples of the unknown vinegar solution, in order to check for precision in the results.

MATERIALS NEEDED

- 25 or 50 mL buret
- 125 mL Erlenmeyer flask
- 5 mL pipet
- pipet bulb
- ringstand
- buret clamp
- commercial vinegar
- standardized NaOH solution
- phenolphthalein solution
- wash bottle
- distilled water

EXPERIMENTAL PROCEDURE

CAUTION: You must use *eye protection* throughout this experiment. The NaOH solution can be harmful to delicate tissues and also to skin with prolonged contact. If NaOH solution gets on your skin or clothing, wash with plenty of cold water.

1. Wash the buret three times with small portions (5–10 mL) of distilled water and then three times with small portions of the standardized NaOH solution. Fill the buret with the NaOH solution and attach it to a ringstand with the buret clamp, as shown in Figure 8.1. Allow some of the solution to run out of the tip of the buret and make sure that there is *no bubble* in the tip. Allow enough solution to run out so that the bottom of the *meniscus* (the curved surface of the solution) is at or below the 0 mark on the buret.
2. Wash a 125 mL flask with tap water and then rinse well with distilled water. Carefully pipet 5.00 mL of vinegar into the flask. You may assume the density of vinegar is 1.00 g/mL, so your unknown sample weighs 5.00 g.
3. Using the wash bottle, add 15–20 mL of distilled water to the flask in such a manner that the water washes the inside surface of the flask. Add three drops of phenolphthalein solution.
4. Make sure that there are no drops of NaOH solution hanging from the tip of the buret. Using the position of the *bottom* of the meniscus, read and record on the data sheet (line d) the initial volume of the NaOH solution to the *nearest hundredth of a milliliter*. A black mark on a sheet of paper held behind and below the meniscus makes the meniscus more visible.

5. Titrate the sample to the *faintest detectable* pink (or violet) color, adding single drops or fractions of a drop of NaOH solution as the end point is approached. (A fraction of a drop may be introduced by manipulating the stopcock to allow a *small* drop to form slowly on the tip of the buret and then touching the drop to the inside of the flask. Again use distilled water to rinse the inside surface of the flask.) Read and record on the data sheet (line c) the final volume of the solution in the buret to the nearest hundredth of a milliliter.
6. Use equation 1 to calculate the number of moles of NaOH used in the titration. This is also the number of moles of CH_3COOH present in the sample. The number of grams of CH_3COOH may be calculated using equation 2, and the percent by weight of CH_3COOH in vinegar may be determined using equation 3.
7. Empty the contents of the 125 mL flask into a sink and wash the flask as described previously. Prepare another sample of vinegar according to steps 2 and 3. Refill the buret and read the initial volume. Calculate the *approximate* number of milliliters of NaOH solution needed to titrate this sample. Add NaOH solution rapidly until the volume added is about 1 mL less than the calculated amount, then carefully complete the titration using single drops and fractions of a drop.
8. Complete at least three titrations, then average the values found for percent acetic acid in the samples.

DATA SHEET

Name: ______________________ Lab Section: _____ Date: __________

TITRATION OF VINEGAR

	# 1	# 2	# 3
a. Mass of vinegar	______	______	______
b. Molarity of NaOH	______	______	______
***c.** Final buret reading	______	______	______
***d.** Initial buret reading	______	______	______
e. Volume of NaOH used (line c − line d)	______	______	______
f. Moles of NaOH used (equation 1)	______	______	______
g. Moles of CH_3COOH present (same as line f)	______	______	______
h. Mass of CH_3COOH (equation 2)	______	______	______
i. Percent CH_3COOH in vinegar (equation 3)	______	______	______
j. Average percent CH_3COOH		______	

* Experimental data

QUESTIONS

1. Why must the buret be rinsed several times with NaOH solution prior to performing the first titration?

2. You are given 135 mL of a 2.00 M solution of CH_3COOH.

   a. How many moles of CH_3COOH are present in the solution?

   b. How many milliliters of a 3.00 M solution of NaOH would be needed to titrate this solution to the end point?

3. How many grams of NaOH could be obtained by complete evaporation of 600 mL of a 5.00 M solution? The formula mass of NaOH is 40.0 g/mole.

EXPERIMENT 9

Heat of Fusion of Ice

OBJECTIVES

1. To become familiar with the calorie as a common unit of heat energy.
2. To learn how transfer of heat energy may be detected and measured.
3. To gain an understanding of the relative amounts of heat energy required to melt ice and to heat water.

PRELABORATORY QUESTIONS

1. What is a calorie?
2. What is meant by fusion?
3. What is a calorimeter?
4. The specific heat of water is 1.00 cal per gram per degree Celsius. What does this mean?
5. The heat of fusion of silver is 21 cal per gram. Explain what this means.

DISCUSSION

The earth is a unique planet in our solar system in that it is the only one whose surface temperature permits large quantities of *water* to exist in the liquid state. Water has a remarkably high capacity to absorb heat energy with relatively little increase in temperature; thus the earth's oceans are vast reservoirs of solar energy and serve to moderate the world's climate.

Water in the solid state, ice, also has a large heat capacity. For example, the amount of heat energy required to melt 1 gram of ice would also be capable of heating 1 gram of water from 20° C (normal room temperature) to 100° C (the boiling temperature).

The melting of a covalent solid such as ice is a *physical change* in which forces of attraction between molecules are broken. This requires an input of heat energy and is therefore an *endothermic* process. All the heat is used in the melting process, so the ice remains at 0° C until it is completely melted; i.e., ice at 0° C melts to water at 0° C. When all the ice has melted, further input of heat energy will raise the temperature of the water produced by the melted ice.

Heat energy is measured in units called *calories.* A calorie is the amount of heat energy needed to raise the temperature of 1 gram of water by 1 Celsius degree.

In this experiment you will determine the *heat of fusion* (melting) of ice using nested styrofoam cups as a simple *calorimeter*. (A calorimeter is an insulated device used for measuring the transfer of heat energy.) A cube of ice will be placed in a measured amount of hot water at a known initial temperature, T_i. As the ice cube melts, it absorbs heat from the hot water, lowering the temperature of the water. Upon melting, the ice cube produces an equivalent mass of water at 0° C. This cold water then absorbs additional heat from the hot water until the hot water and cold water reach a common final temperature, T_f.

The total amount of heat energy H_A given up by the hot water is the *sum* of the heat, H_B, required to melt the ice and the heat, H_C, required to raise the temperature of the ice water to the final temperature of the calorimeter, T_f:

$$H_A = H_B + H_C \tag{1}$$

The heat H_A lost by the hot water can be calculated using the *specific heat* of water, which is the heat energy required to cause a temperature change of 1 degree Celsius for each gram of water. The specific heat of water is 1.00 cal/(g × °C). The temperature *change* in this experiment is $T_i - T_f$, and if g_{H_2O} is the mass of the hot water:

$$H_A = 1.00 \text{ cal/(g} \times {}^\circ\text{C)} \times g_{H_2O} \times (T_i - T_f) \tag{2}$$

In the same manner, the heat H_C gained by the ice water can be calculated using the mass of the melted ice, g_{ice}, its temperature change of $T_f - 0° = T_f$, and the specific heat of water:

$$H_C = 1.00 \text{ cal/(g} \times {}^\circ\text{C)} \times g_{ice} \times T_f \tag{3}$$

The quantity H_B in equation 1 contains the "unknown" *heat of fusion* of ice, H_f, expressed in cal/g:

$$H_B = H_f \times g_{ice} \tag{4}$$

From experimental data, you will be able to determine H_A and H_C using equations 2 and 3. Then H_B can be determined using equation 1 (rearranged):

$$H_B = H_A - H_C \tag{5}$$

When H_B has been determined, the heat of fusion of ice may be calculated by rearranging equation 4:

$$H_f = H_B/g_{ice} \tag{6}$$

MATERIALS NEEDED

- 3 styrofoam cups
- thermometer
- 100 mL graduated cylinder
- 250 mL beaker
- ringstand with ring
- wire gauze
- gas burner
- ice cubes
- towel

EXPERIMENTAL PROCEDURE

1. Set up the ringstand, ring, wire gauze, 250 mL beaker, and gas burner as shown in Figure 9.1. Place approximately 120 mL of water into the 250 mL beaker and heat to 70°C.

   CAUTION: This water is hot enough to cause burns!

   Turn off the burner. Use a cloth-towel to lift the beaker carefully and pour 100 mL of the hot water into the graduated cylinder. Then use the towel to pour the heated water from the graduated cylinder into a styrofoam cup that is nested within another cup (Figure 9.2). This is your calorimeter.
2. Dry an ice cube with a paper towel and place the ice cube in a third cup. Quickly weigh the cup and ice to the nearest 0.1 g and record on the data sheet (line e).
3. Stir the hot water in the calorimeter gently with the thermometer and record the temperature on the data sheet (line b). This is T_i.
4. Without delay, use your fingertips to place the ice cube carefully into the calorimeter. Stir gently with the thermometer until temperature equilibrium is reached, then read and record the temperature on the data sheet (line i). This is T_f. Temperature equilibrium will occur shortly after all the ice has melted.

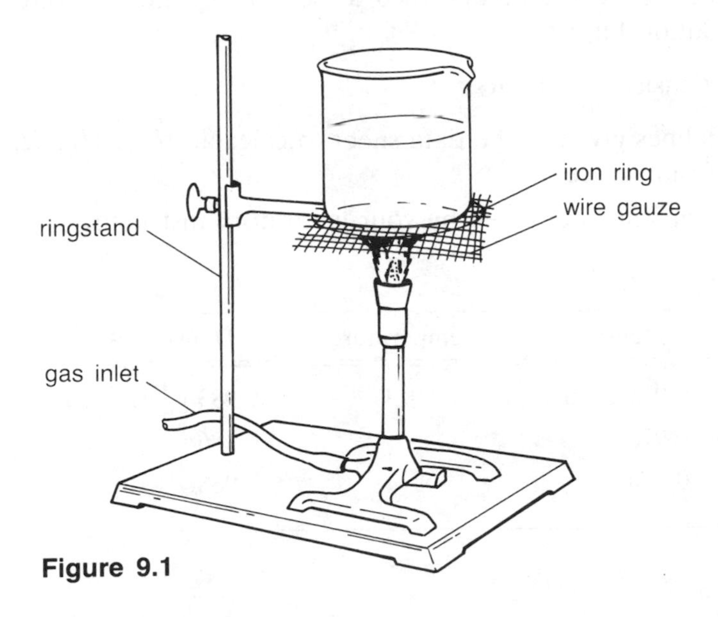

Figure 9.1

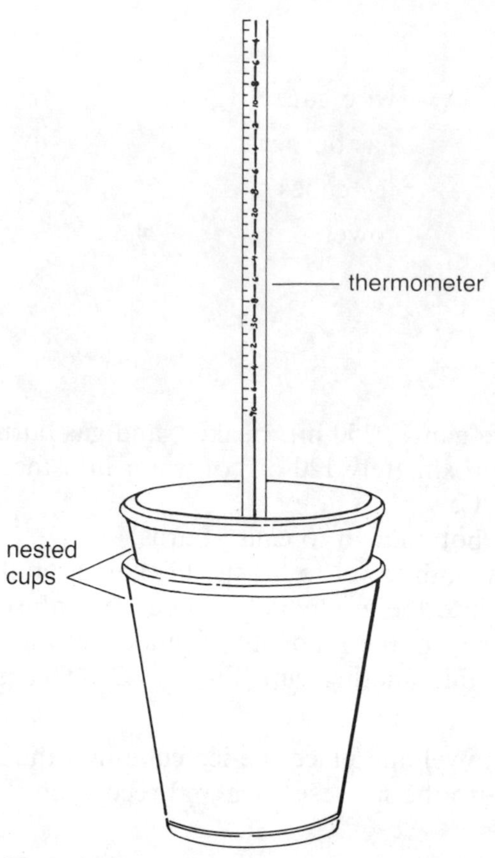

Figure 9.2

5. Reweigh the cup in which the ice cube was weighed and record on the data sheet (line f).
6. The mass of the hot water can be calculated from its density and volume. The density of water depends on its temperature. Use the table given below and the *initial* temperature of your hot water to find its density (values between those given can be estimated). Record this density on the data sheet (line c), the calculate and record the mass of the hot water (line d). The relationship is:

$$g_{H_2O} = \text{density} \times \text{volume} \quad (7)$$

7. Follow the guidelines given in the data sheet to calculate H_A, H_B, H_C, and the heat of fusion of ice, H_f.
8. Repeat the experiment as directed by your laboratory instructor.

DENSITIES OF WATER

Temperature	Density	Temperature	Density
75° C	0.975 g/mL	60° C	0.983 g/mL
70	0.978	55	0.986
65	0.980	50	0.988

DATA SHEET

Name: ______________________ Lab Section: ______ Date: __________

HEAT OF FUSION OF ICE

*a.	Volume of hot water	______	______	______
*b.	Temperature of hot water, T_i	______	______	______
c.	Density of hot water (from density table)	______	______	______
d.	Mass of hot water, g_{H_2O} (equation 7)	______	______	______
*e.	Mass of cup and ice	______	______	______
*f.	Mass of cup	______	______	______
g.	Mass of ice, g_{ice} (line e − line f)	______	______	______
*h.	Initial water temperature, T_i (same as line b)	______	______	______
*i.	Final water temperature, T_f	______	______	______
j.	Temperature change, $T_i - T_f$ (line h − line i)	______	______	______
k.	Calories lost by hot water, H_A (equation 2)	______	______	______
l.	Calories gained by ice water, H_C (equation 3)	______	______	______
m.	Calories needed to melt ice, H_B (equation 5)	______	______	______
n.	Heat of fusion of ice, H_f (equation 6)	______	______	______

* Experimental data

QUESTIONS

1. Assume you would like to prepare a cup of tea. The water from the faucet has a temperature of 20° C. How many calories are needed to heat a cup (236 mL) of this water to 100° C? Assume the density of water is 1.00 g/mL.

2. The heat of fusion of lead is 5.9 cal/g. How much heat energy is required to melt a 20 g piece of lead?

3. Calorimeters used in research have much more insulation than the styrofoam cups used in this experiment. Why is good insulation necessary for accurate calorimetric measurements?

The Molecular Weight of a Volatile Liquid

OBJECTIVES

1. To gain an understanding of the relationships between the pressure, temperature, volume, and mass of a gas.
2. To determine the molecular weight of a substance by use of the ideal gas law.

PRELABORATORY QUESTIONS

1. How would the volume of a mole of gas be affected by:
   a. An increase in temperature (but no change in pressure)?
   b. An increase in pressure (but no change in temperature)?
2. What is the molecular weight of a gas for which a 2 gram sample occupies a volume of 500 mL at a temperature of 7° C and pressure of 700 torr?

DISCUSSION

In our everyday experience we interact constantly with all of the three states of matter: solids, liquids, and gases. The gaseous state is unique in that the molecules are well separated and have very little attraction for one another. These gas molecules are in constant motion. The rate at which they move depends on the *temperature* of the gas. The molecules collide frequently with each other and with the walls of the container. Collisions with the walls of the container result in a force, which is measured as the *pressure* of the gas. Pressure is the force divided by the area on which the force is exerted.

Many years ago it was discovered that the volume (V) of a gas sample is related to both its temperature (T) and pressure (P). These relationships are expressed by simple equations known as the "gas laws." When the number of *moles* (n) of gas present in the sample is also taken into account, these four quantities are related by the *ideal gas law*:

$$PV = nRT \tag{1}$$

In this equation, R is a proportionality constant. The units used for R depend on the units used for V and P. For example, if P is measured in torr and V is measured in milliliters, $R = 6.24 \times 10^4$ (mL $\times$ torr)/(mole $\times$ K). T *must* be expressed in *Kelvin* degrees (K):

$$K = °C + 273 \tag{2}$$

Equation 1 has four variables: P, V, T, and n. If any *three* of these can be measured, the fourth can be calculated from the equation. In this experiment you will measure P, V, and T for a gas sample and calculate the number of moles, n, present in the sample. You will also measure the *mass* (g) of the sample so that the *molecular weight* (MW) of the gas can be determined from the relationship:

$$MW = g/n \tag{3}$$

Your instructor will provide an unidentified liquid with a low boiling point. This will be converted to a gas by heating it in a container open to the atmosphere. Thus the pressure of the gas will be the local atmospheric pressure, and the volume of the gas will be the volume of the container. The temperature and mass of the gas will also be measured so that equations 1–3 can be used to determine the molecular weight of the gas.

To determine the mass of the gas sample, the container (a 125 mL Erlenmeyer flask with a one-hole rubber stopper and capillary tube) is weighed before the sample is introduced and again after the liquid sample is converted to a gas. The gas condenses to a liquid upon cooling and therefore does not readily escape from the flask.

Due to the design of the equipment, one modification in the above procedure is required. The rubber stopper has a tendency to absorb the liquids used as samples. To determine the weight of sample absorbed, the stopper and capillary tube are weighed at the beginning of the experiment and again at the end of the experiment. The difference in these weights is the weight of the absorbed sample. The apparent weight of the gas must be reduced by this amount to obtain the actual weight of the gas.

MATERIALS NEEDED

125 mL Erlenmeyer flask
one-hole rubber stopper with capillary tubing
600 mL beaker
10 cm piece of glass tubing
ringstand with 3 inch and 5 inch rings
clamp
wire gauze
thermometer
gas burner
10 mL graduated cylinder
100 mL graduated cylinder

barometer
"unknown" liquid sample
boiling chips
safety goggles
towel

EXPERIMENTAL PROCEDURE

1. Make sure that the 125 mL Erlenmeyer flask and the rubber stopper with capillary tube are dry. Use an analytical balance to weigh the stopper and capillary tube and record the mass on the data sheet (line a). Then weigh the combined apparatus (flask, stopper, and capillary tube) and record the mass on the data sheet (line b).
2. Set up a ringstand with a 3 inch ring and wire gauze so that the gauze is about 5 cm above the top of a gas burner placed on the base of the ringstand. Place a 5 inch ring about 8 cm above the gauze, and position a clamp on the ringstand above the 5 inch ring, as shown in Figure 10.1. Mount a 10 cm piece of glass tubing in the clamp.
3. Put approximately 400 mL of water into a 600 mL beaker and add several boiling chips. Place the beaker on the gauze and inside the 5 inch ring (Figure 10.1).
4. Your instructor will assign you an "unknown" for this experiment. Record its code number or letter on the data sheet.

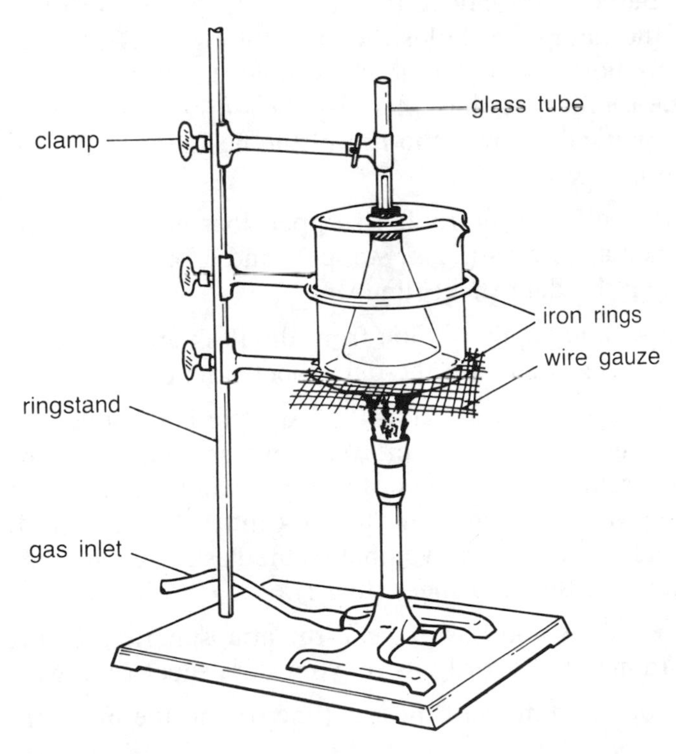

Figure 10.1

CAUTION: The "unknowns" for this experiment were chosen for their low flammabilities. However, they can be *toxic* if inhaled in sufficient quantities or with prolonged contact with the skin. Thus *avoid breathing vapors* of these compounds and *avoid skin contact* whenever possible.

Use a 10 mL graduated cylinder to measure a 5 mL sample of the unknown liquid. Pour the sample into the 125 mL flask and wipe the inside of the neck of the flask with a tissue. Insert the stopper with capillary tube into the neck of the flask.

5. Carefully place the flask into the beaker of water. Slide the glass tubing in the clamp over the capillary tubing. Then use the clamp and tubing to submerge the flask until its bottom is about 1 cm from the bottom of the beaker (see Figure 10.1). (The flask should be submerged up to the stopper. Add water if necessary.) Make sure the clamp is firmly attached to the ringstand.
6. Heat the water to boiling and allow it to boil for several minutes. **CAUTION:** Maintain a safe distance from the apparatus since the water may bubble out of the beaker. Wear safety goggles.
7. Observe the liquid in the flask and in the capillary tubing. The liquid in the flask will boil. Some will condense in the capillary, but this will be forced out by vapor from below. *After all the liquid in the flask has vaporized,* and the liquid (if any) in the capillary remains at a constant level, measure the temperature of the water (this is also the temperature of the gaseous sample). Hold the thermometer so that the bulb is about 5 cm below the surface of the water. Record the temperature on the data sheet (line c).
8. Turn off the gas burner and remove the flask from the beaker of boiling water by raising the clamp that holds the glass tubing. **CAUTION:** The flask and water are hot! Use a *cloth towel* to hold the flask. Immediately dry the top of the capillary tubing, then dry the stopper and the outside of the flask. Dry particularly well the rim of the flask where the stopper is in contact with it.
9. Allow the flask to cool until the rubber stopper does not feel warm, then weigh the combined apparatus (flask, stopper, and capillary tube). Record the mass on the data sheet (line d).
10. Remove the stopper and capillary tube from the flask and weigh immediately. Record this mass on the data sheet (line e).
11. Fill the flask to the *top* with water. Push the stopper and capillary tube into the flask, forcing out excess water. Wipe excess water from the outside of the apparatus, then remove the stopper and capillary tube. Carefully pour the water remaining in the flask into a 100 mL graduated cylinder in order to determine the volume of the flask. Record the volume in milliliters on the data sheet (line f).
12. Your instructor will show you how to read the atmospheric pressure (in torr) from the barometer. Record this on your data sheet (line g).
13. Calculate the uncorrected mass of the gas (line h) and the mass of the

sample absorbed by the stopper (line i). Then calculate the *corrected* mass of the gas (line j).

14. Use the ideal gas law, equation 1, to calculate the number of moles of gas present in your sample (line k). Remember to convert temperature to *Kelvin* degrees using equation 2. Use $R = 6.24 \times 10^4$ (mL × torr)/(mole × K).
15. Use equation 3 to calculate the experimental molecular weight of the gas (line l).
16. Repeat the experiment with another sample of the same unknown.
17. Average the two experimental molecular weights (line m). Your instructor will reveal the formula of the unknown substance so that you can calculate its actual molecular weight (line n). The difference between the actual molecular weight and the average experimental molecular weight is the *error* (line o). Finally, calculate the percent error (line p).

DATA SHEET

Name: ______________________ Lab Section: ______ Date: __________

THE MOLECULAR WEIGHT OF A VOLATILE LIQUID

	Unknown Code	______	
*a.	Mass of stopper and capillary tube	______	______
*b.	Mass of apparatus (flask, stopper, and capillary tube)	______	______
*c.	Temperature of water bath	______	______
*d.	Mass of apparatus and sample (after heating and cooling)	______	______
*e.	Mass of stopper, capillary tube, and absorbed sample	______	______
*f.	Volume of apparatus	______	______
*g.	Atmospheric pressure	______	______
h.	Uncorrected mass of sample (line d − line b)	______	______
i.	Mass of absorbed sample (line e − line a)	______	______
j.	Corrected mass of gaseous sample (line h − line i)	______	______
k.	Moles of gas present (equation 1)	______	______
l.	Experimental molecular weight (equation 3)	______	______
m.	Average molecular weight	______	
n.	Actual molecular weight	______	
o.	Error	______	
p.	Percent error (line o/line n × 100)	______	

* Experimental data

QUESTIONS

1. During step 9 of the experiment, a small amount of liquid often is observed inside the flask even though none was present when the flask was hot. What is this liquid? Will its presence introduce error in the measured mass of the gas? Explain.

2. Explain how the experimental molecular weight would be affected if you did *not* take into account the liquid absorbed by the rubber stopper.

3. It is known that 1.0 mole of a gas occupies a volume of 22.4 L at 0° C and a pressure of 1.0 atmosphere. Determine the value of the gas constant, R, in units of $(L \times atm)/(mole \times K)$.

EXPERIMENT 11

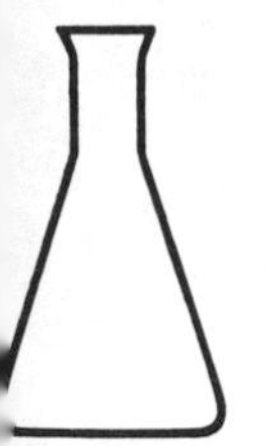

Properties of the Common Acids and Bases

OBJECTIVES

1. To gain experience in selecting the best acid or base for a specific purpose.
2. To learn how, by appropriate dilution, to prepare acids or bases of any strength or concentration.
3. To acquire an appreciation and respect for safe and proper handling of acids and bases.

PRELABORATORY QUESTIONS

1. A lab partner spatters a chemical in his eye. What steps should you take on his behalf?
2. Laboratory bench reagents, because they are in constant use, are commonly contaminated. List four precautions you must take to minimize reagent contamination.
3. Define molarity and normality as these terms apply to chemical solutions.
4. What is an acid? What is a base?

DISCUSSION

It is unfortunate but true that ordinary bench reagents are usually neglected in the study of chemistry. It's easy to overlook the commonplace. After all, the bench reagents are always there and they are constantly in use. For this very reason special care must be exercise to prevent their contamination. *Never* dip a pipet, stirring rod, or test paper into a reagent bottle! Rather, pour only what is needed into a clean receptacle of appropriate size, and dispose of any small excess by pouring cautiously into the sink. *Never* return unused reagent to its bottle!! *Never* place a reagent bottle cap on a bench top in such a way that it might become contaminated, and *always* replace bottle caps promptly and return bottles to their designated place on the shelf.

Each bench reagent is a unique chemical that is always there because of its own distinctive and useful properties. This experiment is a review of the properties of the four "first string" acids and the three "first string" bases—a comparative analysis. (General definitions of acids and bases are given in Experiment 7.) Please think through each of the relatively trivial test tube observations carefully. There is much practical information in this experiment despite its relative simplicity.

Consider these fundamental safety rules:

1. **Chemicals in the eyes:** Always wear splash-proof safety goggles when working with wet chemicals. Regardless of the nature of the chemical, **flush first with copious amounts of water**. Know where the eyewash fountain is and know how to use it! If the irritant is an acid, then place a few drops of castor oil in the eye and seek medical attention. If the irritant is a base, use an eye cup and flush with saturated boric acid solution. Both boric acid solution and castor oil are located in the medicine cabinet. If you wear contact lenses, remove them immediately. Act as quickly as possible—and be prepared to help others who might have an accident.
2. **Chemicals on the clothing or skin:** Flush immediately with lots of water. On the skin, a thorough washing with soap and water is very effective. Acids on clothing can be blotted with a towel well soaked in dilute ammonia (add a few milliliters of NH_3 to a beaker and fill with water). Bases on clothing can be similarly sponged with dilute acetic acid (add a few milliliters of glacial acetic acid to a beaker and dilute with water). An early laundering or dry cleaning may then save the fabric.
3. **Chemicals on the desk top or floor:** Clean up immediately! Sprinkle a strong acid with solid lime, or treat with the 10% sodium carbonate solution immediately. Then mop up before others inadvertently track it about! The adroit handling of a mop is a useful laboratory skill—a must for the careless, the impulsive, the impatient, and the clumsy. Professional mopping calls also for copious amounts of water. **Rinse the mop out after you are through!**

MATERIALS NEEDED

acetic acid, concentrated
ammonia, concentrated
hydrochloric acid, concentrated
nitric acid, concentrated and diluted
sulfuric acid, concentrated
calcium hydroxide solution, saturated
large beakers, 2
small beakers
sodium hydroxide pellets
calcium carbonate, solid chips
copper or silver (small, thin pieces)
benzaldehyde
sucrose, granulated
limestone, powdered, technical grade (for spills)
glass tubing, No. 5, 6, or 7
glass wool

forceps
test tubes, twelve 4 in., four 6 in.
wax pencil
watch glass
litmus paper, red and blue
emery paper strips
aluminum foil
cotton, silk, and wool (white fabric strips)
sodium carbonate solution, 10% (for spills)
safety goggles

EXPERIMENTAL PROCEDURE

The seven reagents scrutinized may be examined in any order. Basic information is provided for each, along with a few suggestions for simple demonstrations of the characteristic chemical properties of each. Please do not feel confined to the illustrations listed. You are encouraged to try others—at least enough to answer the questions confidently. Your answers (or conclusions) will measure your ability to see and understand events in the laboratory. Record all your observations on the data sheet.

It is conventional to consider any dilute reagent to be a 6 normal (6 N) solution if it is not otherwise labeled. Concentrated solutions are labeled as such and have the concentrations under which they are universally marketed. Read the labels carefully before using. Failure to do so is a frequent and dangerous "source of error."

You may well find, as have students before you, that the information provided in this experiment for the common reagents makes an extremely convenient handbook of useful data. (Health science professionals prepare a lot of solutions.)

1. *Sulfuric acid, H_2SO_4 (oil of vitriol):* Concentrated sulfuric acid has a specific gravity of 1.84 and is 96% H_2SO_4 by weight. It is a powerful dehydrating agent and has a heat of solution of −17.75 kcal/mol. It is 18 molar (36 normal) in concentration.
   a. Place approximately 0.1 g of sugar ($C_{12}H_{22}O_{11}$) in a dry test tube and add about a milliliter of concentrated H_2SO_4. Stir and observe.
   b. Place a drop of concentrated H_2SO_4 on a piece of cotton, wool, or silk in a beaker or on a watch glass.
   c. Make a 6 N solution (approximate) of H_2SO_4 by *carefully* adding 1 mL of concentrated H_2SO_4 to 5 mL of water in a test tube. Note the heat effect. Label the test tube and set it aside for use in the last part of this experiment.

2. *Nitric acid, HNO_3 (aqua fortis):* Concentrated HNO_3 has a specific gravity of 1.42 and is about 70% acid by weight. It is a powerful oxidizing agent as well as a strong acid. It is about 16 molar (16 normal). It will readily nitrate many organic materials. A drop of concentrated HNO_3 on a protein material produces a yellow nitrated protein (the xanthoproteic test for proteins).
   a. Place a drop of concentrated HNO_3 on a piece of cotton as well as on a piece of wool or silk, again in a beaker or on a watch glass. After a few seconds, rinse with water and observe the effects.

b. Place a small piece of cotton cloth, $(C_6H_{10}O_5)n$, in a test tube, add concentrated HNO_3, and boil for a couple of minutes. The equation for the reaction that is taking place is:

$$(C_6H_{10}O_5)n + 8n\ HNO_3 \longrightarrow 6n\ CO_2\uparrow + 9n\ H_2O + 8n\ NO\uparrow$$

NO is a colorless gas, but it reacts with oxygen in the cold to give the red-brown gas NO_2.

3. *Hydrochloric acid, HCl (muriatic acid):* Concentrated HCl (38% by weight) has a specific gravity of 1.18. HCl is a gas that is readily soluble in alcohol, ether, and benzene, as well as in water. Concentrated HCl is 12 molar (12 normal).

a. Place a small piece of copper or silver metal in a test tube and add 3 mL of concentrated HCl. Heat gently and observe. *Cool* and carefully add 1mL of concentrated HNO_3. Observe again. This 3 to 1 mixture of concentrated HCl and HNO_3 is called *aqua regia* and is, at the same time, a powerful oxidizing agent and a strong acid. It is so called because even the noble metals (Pt, Au) can be dissolved by it.

b. Using the concentrated HCl on the reagent shelf, prepare 6 mL of 6 N HCl for use in the last part of this experiment. Remember that $mL_1 \times N_1 = mL_2 \times N_2$.

4. *Acetic acid, $HC_2H_3O_2$:* Concentrated acetic acid (glacial acetic acid) is normally marketed as 99.5% pure. Its specific gravity is 1.06. Pure 100% acetic acid is an organic liquid with a boiling point of 118° and a freezing point of 16.7°. Although a relatively weak acid, it is an excellent solvent not only for ionic substances but also for many water-insoluble organic materials. It is readily absorbed into the skin and will cause severe and painful burns as a result. It is palatable when dilute (vinegar is essentially a 3% to 6% acetic acid solution): Indeed, it is a prominent intermediate in the metabolism of digestion products. Concentrated acetic acid is approximately 17.6 molar (17.6 normal) as an acid.

a. Test the solubility of benzaldehyde (artificial oil of almond) in water by adding one drop to 1 mL of water in a small test tube. Test its solubility in glacial acetic acid in the same way.

b. Soak a *small* wad of glass wool in some glacial acetic acid contained in a small beaker (*do not dip anything directly into the reagent bottles*) and, holding the saturated piece of glass wool in a forceps, test to see if the acid is combustible. Try the same test with concentrated HCl and with a fresh wad of glass wool.

c. Prepare 6 mL of 6 N $HC_2H_3O_2$ from glacial acetic acid, label the sample, and save it for use in the last part of this experiment.

5. *Sodium hydroxide, NaOH (caustic soda; lye):* This reagent can be purchased in pellet, flake, or stick form. It is not only very deliquescent, but it also readily absorbs CO_2 from the atmosphere when not stored in a tightly sealed container. Thus it is difficult to maintain in a pure state.

Sodium hydroxide is more destructive of tissues than any of the acids of equal concentration. Proteins are readily digested by it. At room temperature, a saturated solution of NaOH is 52.6% NaOH by weight and has a specific gravity of 1.55. The heat of solution of NaOH is −9.94 kcal/mole. Its properties are similar to those of KOH.

a. Place eight or nine pellets of NaOH in a test tube. Note that they become sticky almost immediately upon exposure to the air. Add about 6 mL of water and note the temperature effect. Label and save half of this solution for the last part of this experiment; use the other half for part b.

b. Place a strip of wool or silk in 3 mL of the NaOH solution and warm gently for a few minutes. Remove with a forceps, rinse thoroughly with water, and test the physical strength of the fabric.

6. *Calcium hydroxide, $Ca(OH)_2$ (soda lime; limewater—when in solution):* A saturated solution of $Ca(OH)_2$ contains only 0.16% base by weight. Insofar as it is soluble, $Ca(OH)_2$ is a strong base. However, its sparing solubility prevents a concentration that is caustic, and it is used quite safely. Of similar strength, $Mg(OH)_2$ can be made into a slurry and taken internally (milk of magnesia) as an antacid and a laxative. Calcium hydroxide has a strange solubility characteristic: It is only about half as soluble in water at boiling temperature as it is in water at room temperature. Calcium hydroxide readily neutralizes all acids, and its solutions—as well as its solid form—absorb CO_2 from the atmosphere to form calcium carbonate. Barium hydroxide ($Ba(OH)_2$) behaves very similarly.

a. Place a few milliliters of limewater in a test tube and, using a piece of glass tubing, blow exhaled air through it for a time. Note the precipitate formed. If you continue to blow air through long enough, the precipitate will dissolve. Try it.

7. *Ammonia, NH_3 (ammonium hydroxide; NH_4OH):* Concentrated ammonia is 59% NH_3 by weight, has a specific gravity of 0.90, and is effectively 15 molar (15 normal). Ammonia is a highly water-soluble gas with a suffocating odor. Compared to most laboratory reagents, however, it is relatively harmless. In addition to being a useful weak base, it is an excellent cleaning agent because it cuts grease so well. By virtue of its structure, it is a most versatile Lewis base, even in the gaseous state.

a. Place about 1 mL of concentrated NH_3 in a test tube. Moisten a strip of red litmus paper and hold it in the fumes, above the mouth of the test tube. Record your observations.

b. Place a drop of concentrated ammonia on the bottom of an upturned beaker. On a second upturned beaker placed next to it, place a drop of concentrated HCl. What do you observe? Explain what happens.

c. Prepare 6 mL of 6 N NH_3 from the concentrated NH_3 on the reagent shelf. (*Remember:* $mL_1 \times N_1 = mL_2 \times N_2$.) Label this sample and save it for the last part of this experiment.

8. *Relative Strengths of Acids and Bases:* Arrange 12 small test tubes, numbered as shown, to contain the following (i.e., two tubes of each):

1 and 7 — the 6 N H_2SO_4 prepared earlier

2 and 8 — 6 N HNO_3 from the reagent shelf

3 and 9 — 6 N HCl prepared earlier

4 and 10 — 6 N $HC_2H_3O_2$ prepared earlier

5 and 11 — dilute NaOH prepared earlier

6 and 12 — 6 N NH_3 prepared earlier

a. In each of the tubes numbered 1 through 6 place a small piece of aluminum foil that has been polished with emery paper. Make note of any reaction that occurs. Try to establish the order of reactivity. If some do not seem to react at all, warm *gently*.

b. In each of the tubes numbered 7 through 12 place a small chunk of limestone ($CaCO_3$). Observe what happens, and again establish the relative order of reactivity.

DATA SHEET

Name: ______________________________ Lab Section: ______ Date: ____________

PROPERTIES OF THE COMMON ACIDS AND BASES

Observations

1. H_2SO_4: **a.** concentrated H_2SO_4 + sucrose: ____________________

____

b. concentrated H_2SO_4 + __________ (wool, silk, or cotton)

____

2. HNO_3: **a.** concentrated HNO_3 + cotton: ____________________

____

concentrated HNO_3 + wool or silk: ____________________

____

3. HCl: **a.** aqua regia + Cu or Ag: ____________________

____

4. $HC_2H_3O_2$: **a.** solubility of benzaldehyde in glacial acetic acid and in water:

____

b. combustibility of glacial acetic acid: ____________________

____

combustibility of concentrated HCl: ____________________

____

5. NaOH: **b.** reaction of hot NaOH solution on wool or silk: __________

____

6. $Ca(OH)_2$: Write the balanced equation for the reaction of $Ca(OH)_2$ with CO_2:

7. NH_3: **a.** ammonia gas + moist litmus: ______________________

____

b. Write an equation for the gas phase reaction you observed:

8. Relative strengths of acids and bases:

a. Order of *decreasing* reactivity with aluminum:

____

b. Order of *decreasing* reactivity with calcium carbonate:

____

QUESTIONS

1. You observed the dehydrating reaction of concentrated H_2SO_4 on sucrose, $(C_{12}H_{22}O_{11})$. Write an equation showing the dehydration of another carbohydrate, cotton cellulose, $(C_6H_{10}O_5)_n$, with hot concentrated H_2SO_4.

2. Calculate the molarity, M, of concentrated hydrochloric acid from the information provided in this experiment for HCl.

3. The sulfuric acid in a fully charged lead storage battery is an aqueous solution 40% H_2SO_4 by weight and has a density of 1.30 g/mL. Calculate the molarity, M, of this battery acid.

4. Why is it observed that ammonia and hydrochloric acid bottles on the shelf frost up so much faster than do sodium hydroxide or sulfuric acid reagent bottles?

5. Suppose that you are running an *organic* reaction, and a sticky brown tarlike material remains in the flask when you are through. This residue does not dissolve in any of the usual solvents, and soap and water will not cut it. Recall your observations of the effect of hot concentrated HNO_3 on cotton, and suggest a way in which you might clean the flask.

6. Write the equation illustrating the absorption of CO_2 gas by limewater.

7. Write the balanced equation for the reaction of aluminum with acetic acid.

8. Ten percent NaOH is a common laboratory reagent. It has a density of 1.111 g/mL. What are its molarity and normality?

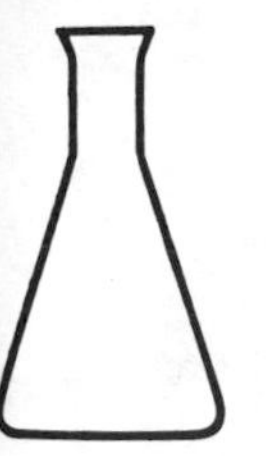

Detection of Some Inorganic Ions in Urine

OBJECTIVE

To demonstrate qualitative tests for common inorganic cations and anions and to use these tests for the analysis of urine.

PRELABORATORY QUESTIONS

1. Why do medical personnel regard the analysis of urine as an important diagnostic procedure?
2. Explain the difference between qualitative analysis and quantitative analysis.
3. Describe how a simple analysis for the presence of sodium ion may be carried out.

DISCUSSION

Urine contains the soluble wastes from metabolic processes of the body. The most abundant of these waste products is urea, an organic compound. Next in abundance are the ions Cl^- and Na^+. Other inorganic ions that may be found in concentrations of 0.01% or more in the urine of healthy individuals include NH_4^+, K^+, Ca^{2+}, PO_4^{3-}, and SO_4^{2-}.

Clinical laboratories run many routine analyses of urine from patients, because the occurrence of unusually high or low concentrations of certain waste products in urine can reveal the presence of disease or organ malfunction. Perhaps the best known of these tests is one that detects glucose, the presence of which may indicate the patient has diabetes mellitus.

Many of the analyses run by clinical laboratories are *quantitative analyses;* i.e., they are designed to determine the *amount* or concentration of a substance. These analyses often require elaborate laboratory equipment. Another type of analysis is *qualitative analysis,* which simply detects the *presence* or *absence* of a substance without regard to its quantity. The qualitative analysis of urine for the inorganic ions listed above can be carried out using simple laboratory equipment.

In this experiment, you will test your own urine for the presence of Na^+, NH_4^+, K^+, PO_4^{3-}, Ca^{2+}, Cl^-, and SO_4^{2-} ions. First, in order to become familiar with the analysis for each of these ions, you will perform the tests on standard solutions known to contain the ion for which the test is designed.

Na^+ may be detected by the characteristic *yellow* color it emits when its electrons are excited by the intense heat of a gas burner. Experimentally, this is accomplished by heating small quantities of the sample on the end of a nichrome wire.

The presence of NH_4^+ is detected by the generation of gaseous ammonia, NH_3, when the sample is treated with strong base:

$$NH_4^+ + OH^- \longrightarrow NH_3\uparrow + H_2O$$

The gaseous NH_3 reacts with moist, red litmus paper, causing it to turn blue.

The remaining ions are identified by the occurrence of precipitates when selected reagents are added:

K^+ is detected by the formation of a white precipitate when sodium tetraphenylborate, abbreviated $NaBPh_4$, is added to the sample:

$$K^+ + NaBPh_4 \longrightarrow \underline{KBPh_4} + Na^+$$

Note that this insoluble potassium compound is a rare exception to the solubility rule given in Experiment 4.

If PO_4^{3-} is present, a yellow precipitate forms when "molybdate reagent" is added to the sample. Molybdate reagent is an acidified solution containing ammonium ions and molybdate ions, MoO_4^{2-}. The reaction is:

$$24\ H^+ + 3\ NH_4^+ + PO_4^{3-} + 12\ MoO_4^{2-} \longrightarrow \underline{(NH_4)_3PO_4 \cdot 12\ MoO_3} + 12\ H_2O$$

Ca^{2+} forms a white precipitate when oxalate ion, $C_2O_4^{2-}$, is added:

$$Ca^{2+} + C_2O_4^{2-} \longrightarrow \underline{CaC_2O_4}$$

Cl^- forms a white precipitate in the presence of Ag^+ ion:

$$Ag^+ + Cl^- \longrightarrow \underline{AgCl}$$

Sulfate ion is detected by its reaction with Ba^{2+} to form white, insoluble $BaSO_4$:

$$Ba^{2+} + SO_4^{2-} \longrightarrow \underline{BaSO_4}$$

MATERIALS NEEDED

ringstand with ring	gas burner	small test tubes
wire gauze	100 mL beaker	nichrome wire
red litmus paper	6 M HNO_3	0.1 M NaCl
Pasteur pipets	6 M NaOH	0.02 M $(NH_4)_2SO_4$
sample bottles	0.1 M $BaCl_2$	0.01 M $Ca(NO_3)_2$

distilled water	0.2 M $(NH_4)_2C_2O_4$	0.02 M K_3PO_4
urine (30 mL)	0.1 M $AgNO_3$	0.02 M $NaBPh_4$
6 M HCl	molybdate reagent	

EXPERIMENTAL PROCEDURE

Collect at least 30 mL of an early-morning sample of urine in a bottle that can be capped. (Early-morning samples have a higher concentration of ions than those taken later in the day.) If possible, refrigerate the sample until your laboratory period begins. Label the bottle with your name, date, and contents.

1. *Sodium ion, Na^+*

a. Clean the nichrome wire to be used for the flame test as follows: Add approximately 2 mL of 6 M HCl to a small test tube. Heat the end of the nichrome wire in the gas burner until it glows, then dip it into the test tube of 6 M HCl. Heat the wire again and dip it into the HCl solution. Repeat this procedure several times until the wire produces a nearly colorless flame.

b. Dip the clean wire into 2 mL of 0.1 M NaCl contained in a small test tube, then heat it in the flame. The resulting intense yellow color is due to the presence of Na^+. Record your observations on the data sheet.

c. Clean the wire again as described above. Use a Pasteur pipet to transfer 2 mL of urine from the sample bottle to a small test tube. Add ten drops of 6 M HCl. Dip the end of the clean wire into the acidified urine, then hold it in the flame. A *bright* yellow flame is a positive test for sodium. (A faint yellow flame indicates traces of sodium as a background contaminant.) Repeat the test if you are not sure of the result. Record your observations on the data sheet.

2. *Potassium ion, K^+*

a. Add five drops of 0.02 M $NaBPh_4$ solution to 2 mL of 0.02 M K_3PO_4 solution contained in a small test tube. The white precipitate indicates the presence of K^+ ion. Record your observations on the data sheet.

b. To 2 mL of urine in a small test tube, add ten drops of 6 M NaOH. This is to remove NH_4^+ ion, which interferes with the test for K^+. Then add five drops of 0.02 M $NaBPh_4$ to the solution. If K^+ is present, a heavy, white precipitate will form. A faint precipitate probably results from remaining traces of NH_4^+. Repeat the test if necessary. Record your observations on the data sheet.

3. *Ammonium ion, NH_4^+*

a. Set up a ringstand, ring, wire gauze, gas burner, and 100 mL beaker as shown in Figure 9.1 of Experiment 9. Half fill the beaker with water and heat to boiling.

b. Add ten drops of 6 M NaOH to 2 mL of 0.02 M $(NH_4)_2SO_4$ solution contained in a small test tube. Wet a piece of red litmus paper with distilled water and place it across the mouth of the test tube. Place the test tube and litmus in the beaker of hot water (it need not actually be boiling at this time).

The litmus paper will develop a *uniform blue color* within a minute or so due to reaction with gaseous NH_3, which is produced by reaction of NaOH with NH_4^+ in the solution (see Discussion). Blue *spots*, which may appear where the paper touches the test tube, are caused by contact of the basic solution with the litmus paper and are *not* a positive test for ammonium ion. Record your observations on the data sheet.

c. To 2 mL of urine in a small test tube, add ten drops of 6 M NaOH. Carry out the analysis using red litmus paper as described above. Again, look for a *uniform* blue color. Repeat the test with another sample of urine if necessary. Record your observations on the data sheet.

4. *Phosphate ion,* PO_4^{3-}

a. To 1 mL of 0.02 M K_3PO_4 in a small test tube, add 1 mL of molybdate reagent. The yellow precipitate, which develops slowly, indicates the presence of phosphate ion. Record your observations on the data sheet.

b. To 1 mL of urine in a small test tube, add 1 mL of molybdate reagent. Look for the yellow precipitate that is a positive test for PO_4^{3-}. (A greenish precipitate indicates that insufficient molybdate reagent was added.) Repeat the test, if necessary, using another sample of urine. Record your observations on the data sheet.

5. *Calcium,* Ca^{2+}

a. Add ten drops of 0.2 M ammonium oxalate, $(NH_4)_2C_2O_4$, to 2 mL of 0.01 M $Ca(NO_3)_2$ solution contained in a small test tube. The presence of calcium ion causes formation of a fine white CaC_2O_4 precipitate. The precipitate forms slowly; allow the solution to stand for several minutes. Record your observations on the data sheet.

b. To 2 mL of urine in a small test tube, add ten drops of 0.2 M $(NH_4)_2C_2O_4$. Allow the solution to stand for several minutes. Watch for the white precipitate that will form slowly if Ca^{2+} is present. The white color may be masked if the urine has a deep yellow color. In this case, a cloudiness that develops indicates the presence of Ca^{2+}. Repeat the test if necessary. Record your observations on the data sheet.

6. *Chloride ion,* Cl^-

CAUTION: $AgNO_3$ solution may cause dark spots upon contact with your skin. However, these spots are temporary.

a. Add five drops of 0.1 M $AgNO_3$ solution to 2 mL of 0.1 M NaCl. The white AgCl precipitate shows that Cl^- ion is present. Record your observations on the data sheet.

b. To 2 mL of urine in a small test tube, add ten drops of 6 M HNO_3. (This is to prevent interference of PO_4^{3-} ion with the test for Cl^-.) Then add five drops of 0.1 M $AgNO_3$ to the solution. If Cl^- is present, the white AgCl precipitate will form. Repeat the test if necessary. Record your observations on the data sheet.

7. *Sulfate ion, SO_4^{2-}*

a. Add ten drops of 0.1 M $BaCl_2$ solution to 2 mL of 0.02 M $(NH_4)_2SO_4$ solution contained in a small test tube. The white $BaSO_4$ precipitate that forms indicates the presence of sulfate ion. Record your observations on the data sheet.

b. To 2 mL of urine in a small test tube, add five drops of 6 M HCl. (This prevents interference of PO_4^{3-} ion with the test for SO_4^{2-}.) Then add ten drops of 0.1 M $BaCl_2$. Look for the white precipitate indicative of the presence of SO_4^{2-}. Repeat the test if necessary. Record your observations on the data sheet.

DATA SHEET

Name: ______________________ Lab Section: ______ Date: ____________

DETECTION OF SOME INORGANIC IONS IN URINE

1. *Test for sodium ion, Na^+*

b. NaCl solution. Observations:

c. Urine sample. Observations:

Is Na^+ present in the urine sample?

2. *Test for potassium ion, K^+*

a. K_3PO_4 solution. Observations:

b. Urine sample. Observations:

Is K^+ present in the urine sample?

3. *Test for ammonium ion, NH_4^+*

b. $(NH_4)_2SO_4$ solution. Observations:

c. Urine sample. Observations:

Is NH_4^+ present in the urine sample?

4. *Test for phosphate ion,* PO_4^{3-}

a. K_3PO_4 solution. Observations:

b. Urine sample. Observations:

Is PO_4^{3-} present in the urine sample?

5. *Test for calcium ion,* Ca^{2+}

a. $Ca(NO_3)_2$ solution. Observations:

b. Urine sample. Observations:

Is Ca^{2+} present in the urine sample?

6. *Test for chloride ion, Cl^-*

a. NaCl solution. Observations:

b. Urine sample. Observations:

Is Cl^- present in the urine sample?

7. *Test for sulfate ion, SO_4^{2-}*

a. $(NH_4)_2 SO_4$ solution. Observations:

b. Urine sample. Observations:

Is SO_4^{2-} present in the urine sample?

QUESTIONS

1. Describe how you could quickly distinguish between the following pairs of solutions:

   a. $NaCl$ and NH_4Cl

   b. KCl and K_3PO_4

   c. $Ca(NO_3)_2$ and KNO_3

   d. $(NH_4)_2C_2O_4$ and $NaNO_3$

   e. $AgNO_3$ and $NaCl$

   f. $BaCl_2$ and $(NH_4)_2SO_4$

2. When carrying out the analysis for K^+ in urine, the urine is made basic by adding NaOH prior to adding $NaBPh_4$. NaOH prevents interference by the NH_4^+ ion, which is normally present in urine and which also forms a precipitate with BPh_4^-. Explain how the presence of NaOH prevents the NH_4^+ ion from interfering with the K^+ analysis. Write a balanced equation to support your explanation.

3. The *quantitative* analysis for an ion often involves collecting and weighing a precipitate of known composition. Suppose excess $BaCl_2$ solution is added to a 100 mL sample of acidified urine, and a $BaSO_4$ precipitate weighing 0.450 g is collected. What is the molar concentration of SO_4^{2-} in the urine?

EXPERIMENT 13

Osmosis and Dialysis

OBJECTIVES

1. To learn to use dialysis as a purification technique.
2. To understand the physical basis for and mechanism of osmosis.
3. To know the distinction between osmosis and dialysis.

PRELABORATORY QUESTIONS

1. Explain how dialysis and osmosis differ.
2. What is a semipermeable membrane?
3. What happens to a liquid when it is distilled?
4. How does an anion differ from a cation?

DISCUSSION

The walls of living cells are semipermeable membranes. Nutrients and metabolic wastes are allowed to filter through these walls, while the proteins of the cytoplasm and body fluids are denied passage. An ultrafiltration is involved, but the phenomenon is more complicated than that. Among other things, a balance of osmotic pressures is simultaneously in operation, a balance that, when upset even slightly, brings on pronounced physiological changes. You have read in novels of the sea of the grave consequences of drinking salt water. You may have witnessed the severe tissue edema that accompanies certain kidney malfunctions. In the one case the osmotic pressure of the body fluids becomes too high; in the other, too low.

This experiment provides an opportunity to study and compare osmosis and dialysis, two phenomena of great importance in cell physiology.

Dialysis is a form of ultrafiltration (filtration of particles of molecular size) through a semipermeable membrane. The pores of a dialysis membrane are of such a size that solvent and small solute molecules pass through, while large polymeric molecules of colloidal dimensions are restrained. You might ask, then, why one could not simply use a large round piece of dialysis membrane in a Büchner funnel and then filter the colloids from the small solute molecules with suction. Unfortunately, such a procedure will not work. As the large molecules are drawn into position over the holes in the membrane, they are held there by the suction. Ultimately, the pores become so plugged up that even solvent is denied passage. Once impermeable, the membrane either breaks or filtration stops.

Think, if you will, at the molecular level and see what actually happens during *dialysis*, and you will understand why clogging of the membrane pores is not a problem in dialysis (or osmosis). Inside the dialysis bag, molecules of all sizes are moving about freely. The small molecules, if they score a direct hit on a pore, pass through to the other side, where an irrigating stream of solvent sweeps them away before they have an opportunity to wander back through a neighboring pore into the bag. Particles of colloidal size, however, bounce off the membrane walls. There is no force to hold them in position over the pores they encounter. Solvent is flowing in from the outside as freely as solvent from the inside is flowing out. It is as though the membrane surface were being continually swept free of obstructions. We give the molecules an option, apply no force, and the molecules very efficiently separate themselves.

In *osmosis*, the membrane pores are even smaller—so small that liquid flow as such is not possible. The same surface tension that prevents water from soaking through the feathers of a duck is in operation. A liquid surface effectively seals off both the inlet and the outlet of each pore. However, *individual water molecules* can pass through the pores. The process is one of evaporation. Water has a vapor pressure, and individual molecules escape from the liquid surface into the pore of the membrane as gas. These gas molecules can then negotiate the pore space and condense on the liquid surface blocking the other entrance to the pore. Indeed, the osmosis phenomenon may be regarded as a *distillation of solvent* through the pores of a semipermeable membrane. Osmosis is *not* a flow of solution through the pores as is the case in dialysis. A little reflection will justify the statement: In osmosis solvent distills through the pores of a semipermeable membrane from the solution of higher vapor pressure (lower solute concentration) to the solution of lower vapor pressure (higher solute concentration). Note that it is the solvent, not the solution, that passes through the membrane pores during osmosis. Osmosis ceases when the vapor pressures (and effective solute concentrations) on both sides of the membrane are equal. This state of equilibrium is often referred to as *osmotic balance*.

Consider what takes place when a crystal of $FeCl_3$ is dropped into a solution of sodium metasilicate, Na_2SiO_3 (also known as water glass). As the iron chloride dissolves, the Fe^{3+} ions diffuse from the crystal surface to mingle with the Na^+ and SiO_3^{2-} of the water glass. The Fe^{3+} reacts with SiO_3^{2-} to form an iron metasilicate, $Fe_2(SiO_3)_3$, a water-insoluble precipitate that happens to be structured as a semipermeable membrane. Shortly, an iron metasilicate membrane encloses the remaining $FeCl_3$ crystal along with some of the water, Na^+ ions, and Cl^- ions. Now, within this capsule, $FeCl_3$ continues to dissolve until the concentration of ions within the capsule exceeds the concentration of ions on the outside. An osmotic pressure develops and water proceeds to distill from the external solution through the iron metasilicate membrane into the capsule, until finally the capsule bursts (in its weakest spot). The free Fe^{+3} ions spill through the rupture, a new metasilicate membrane begins to form, and the process repeats itself. Watch the growth of the membrane carefully. It is a graphic illustration of what happens in hemolysis and plasmolysis.

MATERIALS NEEDED

- Benedict's reagent
- iodine/KI reagent
- 6 M nitric acid
- 2% silver nitrate
- sodium silicate (water glass) (20 mL)
- starch/sugar/salt/methylene blue solution (25 mL)
- crystals of soluble heavy metal salts (Cu^{2+}, Co^{2+}, Fe^{3+}, Ni^{2+}, Mn^{2+}, Cr^{3+}, or Zn^{2+}—a small crystal of at least three of these representative salts)
- 50 mL beaker
- 250 mL filter flask (with one-hole stopper and rubber connecting tubing)
- funnel
- test tubes to fit centrifuge
- glass tubing, 12 in.
- Visking tubing, 12 in.
- centrifuge

EXPERIMENTAL PROCEDURE

Obtain 12 inches of Visking tubing, a 250 mL filter flask, and the stopper and rubber tubing necessary to set up an apparatus as shown in Figure 13.1. The

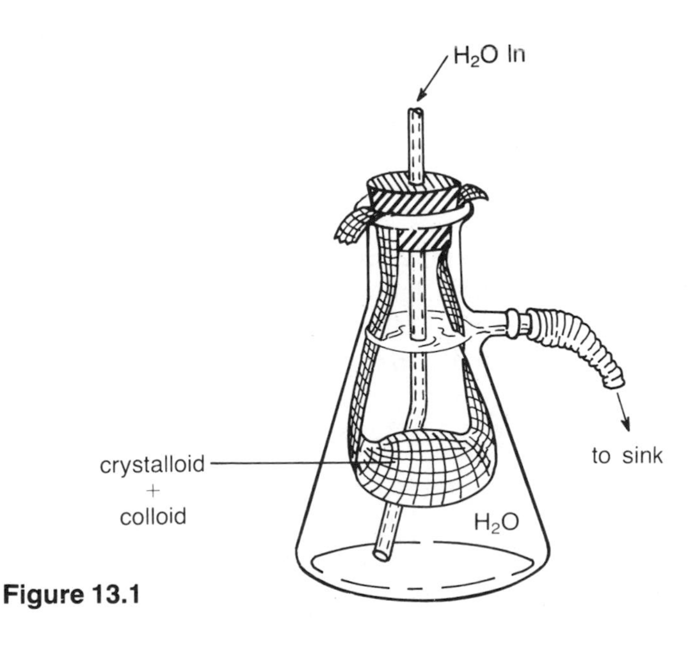

Figure 13.1

impure starch solution, the water glass, the heavy metal salts, and the centrifuges will be found in the laboratory.

Since this is an experiment involving more reflection than manipulation, it is suggested that you team up with one other student and that one of you sets up the dialysis (part A) while the other prepares for the silicate garden (part B). Work together toward understanding the phenomena, fill in the data sheet, and answer the questions.

A. Dialysis; The Purification of a Starch Solution

Set up an apparatus like that of Figure 13.1 but without the Visking dialysis tube. Soak one end of the dialysis tubing in water for a few moments and then open the end by gently rolling the moistened tubing between the thumb and forefinger. Loosen in this manner until about two-thirds of the tubing is open, and then hold both ends at the same level while you introduce 20 mL of the impure starch solution. Use a funnel and have your partner assist if necessary. Secure the bag in the 250 mL filter flask as illustrated in Figure 13.1. Adjust a *gentle* flow of water through the flask and allow the irrigation to proceed until about 15 minutes remain in the laboratory period. (Complete purification usually requires a longer dialysis time. Your instructor may arrange to have a couple of dialyses left on overnight in order to demonstrate the fact.)

Unless you have been told otherwise, your starch solution contains three easily detectable impurities: sodium chloride, glucose, and methylene blue. All are crystalloids, but the latter is relatively large, almost big enough to be colloidal in size. Methylene blue has the molecular formula $(C_{16}H_{18}N_3 \cdot 3H_2O)^+Cl^-$ and a formula weight of 374. It is commonly used as an indicator and as a bacteriological stain.

Examine a few milliliters of the original impure starch solution and note its color. Centrifuge 2 mL of this solution for about 5 minutes. *Be sure that you balance the centrifuge with a tube equally filled with water*. Make note of your observations on the data sheet.

Test a small portion of the centrifuged starch solution for chloride ion by adding a drop of silver nitrate solution. Acidify with a few drops of nitric acid, mix, and see if the formed precipitate dissolves.

To a small volume of the centrifuged starch solution add a drop of iodine solution. A deep blue complex forms if starch is present.

Test the starch solution for the presence of glucose as follows: First heat 2 mL of Benedict's reagent in a test tube for a couple of minutes in a boiling water bath. To this hot reagent, add five drops of the starch solution, mix thoroughly, and heat for an additional 5 minutes in the water bath. If glucose is present, a color change ranging from the original blue (Cu^{2+}) through green (Cu^{2+} + $Cu_2(OH)_2$) and yellow ($Cu_2(OH)_2$) to a brick red (Cu_2O) may be observed. The ultimate color observed depends on the concentration of glucose present.

You will make these same observations and tests on the dialyzed starch at the end of the lab period.

B. Osmosis; The Preparation of a Silicate Garden

Add 20 mL of water glass (sodium metasilicate solution) to 20 mL of water in a 50 mL beaker. Stir well and allow to stand quietly for a few minutes. Drop into this metasilicate solution a single crystal of one or more of the following salts: cupric sulfate, cobalt chloride, ferric sulfate, nickel sulfate, manganese chloride, chromium nitrate, or zinc sulfate. Watch closely as the garden grows, and correlate your observations with what you have learned about the chemical and osmotic driving forces behind the growth.

Water glass actually contains several molecular species. To simplify your answers on the data sheet, assume that the prominent component, sodium metasilicate, is the active ingredient.

DATA SHEET

Name: ______________________ Lab Section: ______ Date: __________

OSMOSIS AND DIALYSIS

A. Dialysis

Describe the appearance of the starch solution or test results in the blank spaces provided.

Test or Observation	Before Dialysis	After Dialysis	Conclusion(s)
Color	______	______	______
Centrifugation	______	______	______
$AgNO_3$ for Cl^-	______	______	______
I_2 for starch	______	______	______
Benedict's reagent	______	______	______

Equation for the silver nitrate test for chloride ion (write both the total ionic equation and the net ionic equation):

The formula weight of the methylene blue *cation* is: ______________________

The *ions* to which the Visking dialysis membrane was permeable are: ______

____

The *molecules* to which the Visking membrane was permeable are: ______

____

B. OSMOSIS

Describe, in your own words, the growing silicate garden.

Write an ionic equation that illustrates the reaction of sodium metasilicate with $Cr(NO_3)_3$.

Explain, at the molecular level, why the silicate sacs of the growing chemical plant almost always burst at the top.

QUESTIONS

1. A solution of ferrous ammonium sulfate is prepared by dissolving 50.0 g of $Fe(NH_4)_2(SO_4)_2 \cdot 6H_2O$ in enough water to make 250 mL of solution. The molecular weight of this ferrous ammonium sulfate is 392. The density of the solution prepared was 1.154 g/mL. Calculate the molarity, M, of this solution.

2. You could have shown, in this experiment, that as long as any methylene blue cation remained in the dialysis tube, Cl^- ion was also present. Quite obviously, the small compact Cl^- anion should have a much greater mobility than the big and cumbersome methylene blue cation. Explain why the rate of cation loss through the membrane is always equal to the rate of anion loss.

3. van't Hoff derived an equation for calculating the osmotic pressure, π, of solutions that looks very much like the familiar universal gas law equation. It is

   $\pi V = nRT$, where:
   - π is the osmotic pressure in atmospheres
   - V is the volume of the solution in liters
   - n is the number of moles of solute
   - R is the universal gas constant, 0.082 liter-atmos/mole-degree
   - T is the absolute temperature of the solution

   Calculate the osmotic pressure (at room temperature of 27°) of a solution prepared by dissolving 18 g of glucose (MW = 180) in enough water to make 1 liter of solution.

4. What would happen to the silicate garden if the "water glass" solution were stirred during its growth?

EXPERIMENT 14

Covalence and Structure: The Use of Molecular Models

OBJECTIVES

1. To provide practice in interconverting two-dimensional structural formulas and three-dimensional models.
2. To illustrate differences between structural, geometric, and conformational isomers.
3. To point out distinctions between molecular, structural, and electronic formulas, and to be able to predict one from another.
4. To provide practice in picturing the structures attendant upon sp^3, sp^2, and sp hybridization.

PRELABORATORY QUESTIONS

1. How many covalent bonds (single bonds) are associated with carbon? With hydrogen? With oxygen? With nitrogen? With chlorine? With sulfur?
2. What is meant by "a saturated hydrocarbon"?
3. What bond angle is normally associated with sp^3 hybridization? With sp^2 hybridization? With sp hybridization?
4. What is a structural isomer?
5. What is a functional group?

DISCUSSION

Select almost any topic fundamental to an understanding of life and health, and it will be based on organic chemistry. Carbohydrates, lipids, proteins, vitamins, hormones, and drugs are all numbered among the approximately eight million known organic compounds in existence. About two million of these have known structures, but the list of characterized organic compounds grows daily.

We are fortunate that this overwhelming host of organic compounds fits within a simple system of classification—a system based on *structure*. Organic structure, in turn, is determined, because of the three hybrid variations of covalence (sp, sp^2, and sp^3), by three basic bond angles.

The organic chemist is a molecular architect. He knows that once he has a clear three-dimensional "molecule's eye view" of a molecule, he will be able to predict its physical properties and chemical behavior. He has a system, and his system is based on structure.

You are urged to take model study seriously. It is meant to provide a perceptual insight to bonding and structure. To learn organic chemistry by rote memorization is totally frustrating if not impossible. Systematized, the discipline can be quite easy and enjoyable.

You are urged also to review the text and classroom materials on bonding and structure before beginning this experiment. Space limitations here prohibit more than an outline of the underlying theory.

Use of the Models

Many different model kits are available. The following guidelines apply to the Sargent Ball and Stick kits but usually can be applied to others as well.

1. A pair of shared electrons constitutes a single covalent bond, represented by a stick in the models. Two electron pairs constitute a double bond, which is usually represented by a pair of springs or by two bent sticks.
2. The normal unstrained angle between sp^3 carbon bonds is 109° 28′. Whenever sp^3 bond angles differ appreciably from this value, a structural strain arises and springs must be used instead of pegs. Sp^3 hybridization exists for any carbon bonded to four other atoms.
3. These models are not built to scale; i.e., the atomic diameters and the bond lengths are of the same dimensions regardless of atoms designated. Actually large differences can exist in the dimensions of atoms. Your instructor may have available a set of more accurate (and more expensive) molecular models. These are usually constructed to a scale where 1 angstrom (10^{-8} cm) is represented by 1 cm. The atoms and the bond lengths in these models are calculated to present a true picture of the external shape of the molecule.
4. The following color identification scheme is appropriate for the Sargent atomic models:

C: black	O: red	Cl: green	I: purple
H: yellow	N: blue	Br: orange	

5. Use the 2 inch pegs for carbon-to-carbon bonds and the short 1 inch pegs for other bonds.
6. Each ball has a number of holes corresponding to its valence except for nitrogen, which has five holes. The two extra holes make possible the construction of typical nitrogen compounds with coordinate covalent bonds as well as covalent bonds.

Kinds of Formulas

The formula is all-important in organic chemistry, but the ordinary molecular formula often does not provide enough information. Consider the simple organic halide, dichloroethane, (CH_2ClCH_2Cl), and the various representative formulas:

Empirical Formula Shows only the elements and their ratio; molecular weight and structure remain unknown.

CH_2Cl

Molecular Formula Shows the elemental composition, ratio of atoms, and molecular weight, but tells nothing about structure.

$C_2H_4Cl_2$

Structural Formula Shows everything the molecular formula does as well as the manner in which atoms are joined to one another (connectivity). It still does not, however, tell us the electron distribution or the conformational arrangements possible. The structural formula shows that there are two isomers of dichloroethane:

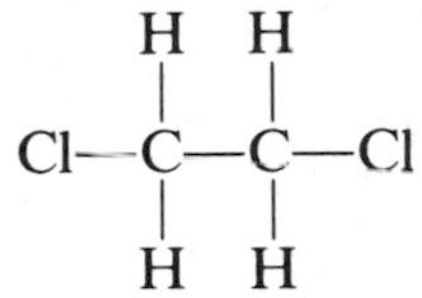

$$
\begin{array}{ccccccc}
 & & H & & H & & \\
 & & | & & | & & \\
Cl & — & C & — & C & — & Cl \\
 & & | & & | & & \\
 & & H & & H & &
\end{array}
\quad \text{or} \quad CH_2ClCH_2Cl
\qquad\qquad
\begin{array}{ccccccc}
 & & Cl & & H & & \\
 & & | & & | & & \\
Cl & — & C & — & C & — & H \\
 & & | & & | & & \\
 & & H & & H & &
\end{array}
\quad \text{or} \quad CHCl_2CH_3
$$

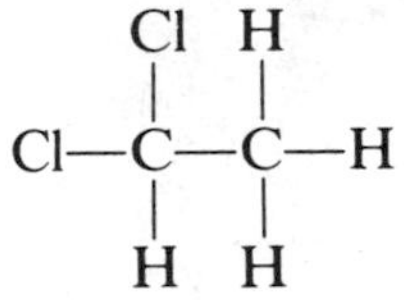

1,2-Dichloroethane   1,1-Dichloroethane

Electronic Formula Includes the distribution of electrons within the structure, but still doesn't distinguish between possible conformations.

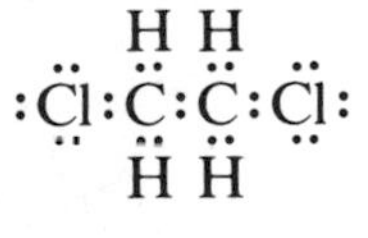

$$
\begin{array}{c}
\quad H\;H \\
:\ddot{\underset{..}{Cl}}:\ddot{\underset{..}{C}}:\ddot{\underset{..}{C}}:\ddot{\underset{..}{Cl}}: \\
\quad H\;H
\end{array}
\quad \text{and} \quad
\begin{array}{c}
\quad H\;H \\
:\ddot{\underset{..}{Cl}}:\ddot{\underset{..}{C}}:\ddot{\underset{..}{C}}:H: \\
:\underset{..}{Cl}:H
\end{array}
$$

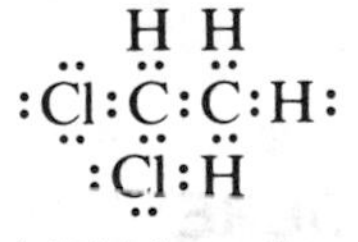

1,2-Dichloroethane   1,1-Dichloroethane

Conformational Formula Shows connectivity (permanent structure) as well as temporary, or conformational, structure. Conformations are structures that are relatively temporary and that can be converted from one to another by rotations about single bonds. For example, the three possible conformations of 1,2-dichloroethane are shown in Figure 14.1.

None of the structures in Figure 14.1 is drawn to scale. It is good to remember also that elements vary in size. Consider the scale model structures of 1,2-dichloroethane in Figure 14.2, and the steric crowding and relative stabilities of the conformations become much more understandable.

MATERIALS NEEDED

ball and stick model sets (one set per student)

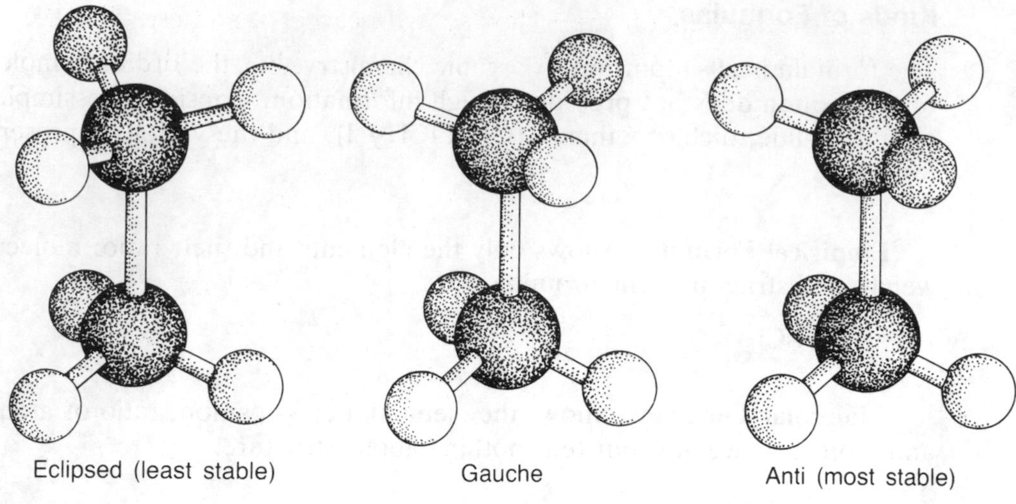

Figure 14.1

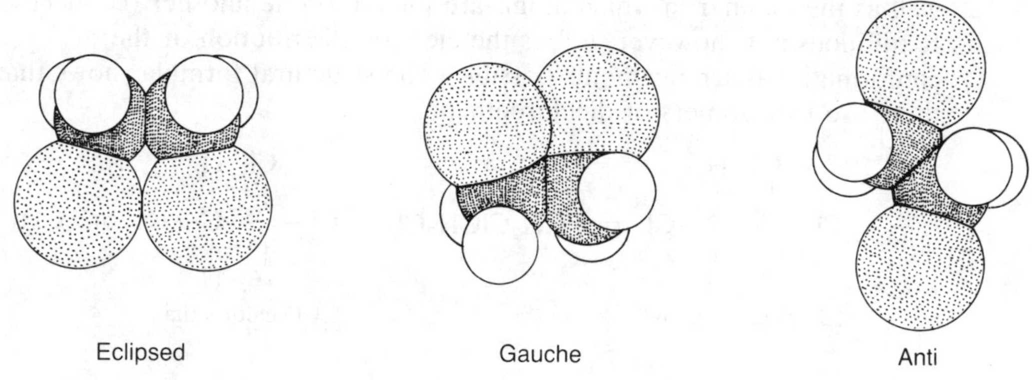

Figure 14.2

EXPERIMENTAL PROCEDURE

Using the models, proceed through the following experiments, answering the questions on the data sheet. In the laboratory, let the instructor verify your structures as you complete them.

1. Prepare and compare structures for:
  a. methane; CH_4 (a gaseous alkane)
  b. methyl iodide; CH_3I (a low-boiling liquid alkyl halide)
  c. methyl alcohol; CH_3OH (an alcohol that boils at 67°)
  d. methyl amine; CH_3NH_2 (a 1° amine; an organic base)
  e. dimethyl amine; $(CH_3)_2NH$ (a 2° amine; an organic base)
  f. trimethyl amine; $(CH_3)_3N$ (a 3° amine; an organic base)
  g. acetone; $CH_3\underset{\underset{\displaystyle O}{\|}}{C}CH_3$ (a ketone; dimethyl ketone)

Answer question 1 on the data sheet.

2. Construct models for the following hydrocarbons, and establish the geometric characteristics of sp^3, sp^2, and sp hybrid bonding:
   a. *n*-propane (a straight chain alkane; sp^3 hybrid bonds)
   b. propene (an alkene; sp^2 hybrid bond)
   c. propyne (an alkyne; sp hybrid bond)
   d. cyclopropane (a cycloalkane; strained sp^3 hybridization)

   Answer question 2 on the data sheet.

3. Construct a model of *n*-hexane and adjust it to its least crowded conformation (all H's as far from one another as possible).

   Convert the *n*-hexane to cyclohexane. Do this by removing a hydrogen from the number one carbon and from the number six carbon and connecting these carbons to make the six-membered ring. You must rotate bonds to accomplish this. Note that two conformations are possible, as illustrated in Figure 14.3.

   Answer question 3 on the data sheet. (*Hint*: First determine all the possible isomeric hexanes and then substitute chlorines for the *different* H's on each.)

4. Note that there are three isomeric dichloroethenes. See Figure 14.4.

   Build structures for the five isomeric compounds having the molecular formula C_4H_8. (*Hint*: There are six isomers of C_4H_8.)

   Answer question 4 on the data sheet.

5. Construct a benzene molecule: C_6H_6,

   Answer question 5 on the data sheet.

   Please put the model kits back in order and return them.

Figure 14.3

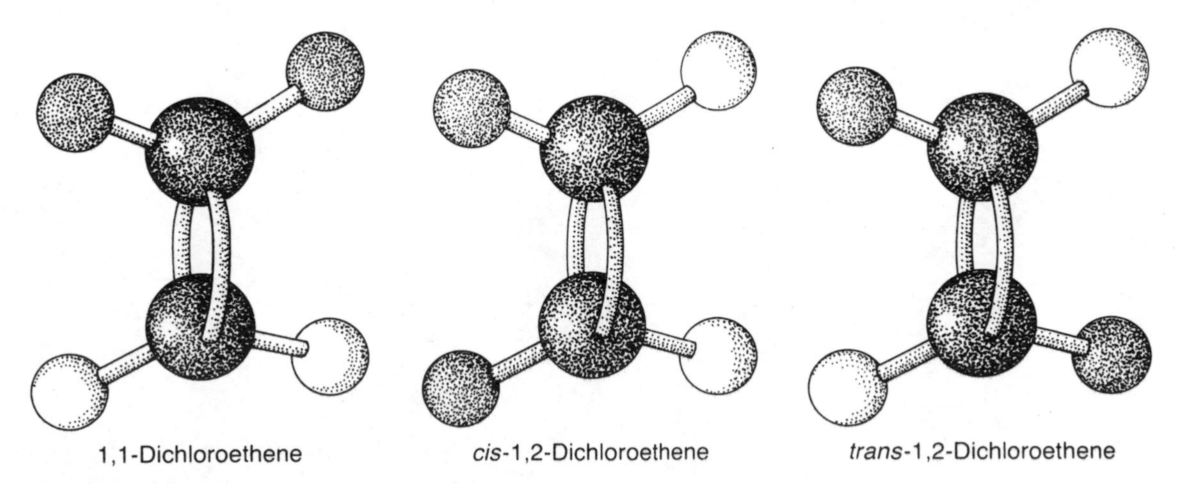

Figure 14.4

DATA SHEET

Name: ______________________ Lab Section: ________ Date: ________

COVALENCE AND STRUCTURE: THE USE OF MOLECULAR MODELS

1. Derivatives of methane

a. Let *R* represent any alkyl group including methyl- and write general formulas for the following:

Alkanes ________ Alkyl halides ________

Alcohols ________ 1° amines ________

2° amines ________ 3° amines ________

Ketones ________

b. Why are the amines basic? (Compare them with their parent, ammonia.)

c. Draw the electronic structure for acetone.

2. Hydrocarbons containing three carbons

a. Which of these three-carbon hydrocarbons are isomers?

b. Draw a structure for propene that illustrates the nature and location of its sigma and pi bonds.

3. *n*-hexane and cyclohexane

a. Write structures and names for all of the isomeric monochloro derivatives of the hexanes—i.e., all of the isomeric $C_6H_{13}Cl$

compounds. (There are 17 of them. Do not list additional conformational isomers, only structural isomers.) Use an extra sheet of paper if necessary.

b. Cyclopentane, as contrasted to cyclohexane, is a nearly flat (planar) molecule. Why is this true?

4. Isomers of molecular formula C_4H_8

a. Draw structures for the six isomers of molecular formula C_4H_8 (name each structure):

5. Benzene

a. Explain why benzene has a single, symmetrical and planar structure, whereas cyclohexane is nonplanar and exists in more than one conformation. (Note bonding hybridizations.)

b. In the presence of a very strong acid such as 100% sulfuric acid, benzene functions as a weak base. Consider its bond hybridization, recall ammonia and the amines, and venture an explanation for its weak but demonstrable basicity.

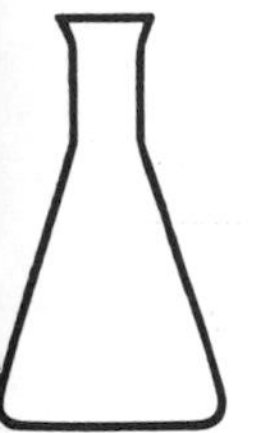

The Preparation and Comparison of Methane and Acetylene

OBJECTIVES

1. To gain experience in handling organic gases.
2. To learn how to distinguish between saturated and unsaturated organic compounds in the laboratory.
3. To learn how to predict products and write equations for the reactions of the test reagents used in this experiment on hydrocarbons of known structure.

PRELABORATORY QUESTIONS

1. What is the difference between a substitution reaction and an addition reaction?
2. Characterize methane and acetylene as to their solubility in water and in organic solvents such as petroleum ether or carbon tetrachloride.
3. What is the difference between a saturated and an unsaturated hydrocarbon?

DISCUSSION

By now you understand the conventional approach to the study of organic chemistry. Its compounds are classified according to structure, and then each class is scrutinized for physical and chemical properties characteristic of its members.

You will investigate methane and, from its chemical behavior, deduce properties common to all alkanes. A similar look at acetylene will help you to discover the characteristics of the alkynes. Several useful test reagents are introduced. Pay careful attention to the relation between structure and chemical reactivity, and organic chemistry will soon begin to transcend the mysterious.

In this experiment we shall synthesize two organic gases and use them as examples to illustrate the differences in chemical behavior between *saturated* and *unsaturated* compounds. The simple chemical tests used may be of value to you in identifying the unknowns in Experiments 16 and 20. These same reactions are also encountered in Experiments 17 and 19.

Methane, CH_4, is the simplest saturated hydrocarbon. Perhaps the easiest method for preparing pure methane involves the decarboxylation of acetic acid or of one of its salts. You will decarboxylate sodium acetate by heating it with soda lime, a mixture of sodium hydroxide and calcium oxide:

$$CH_3C(=O) - ONa + NaOH + CaO \xrightarrow{\text{heat}} CH_4\uparrow + Na_2CO_3 + CaO$$

The CaO serves as a dehydrating agent and expedites mixing of reactants.

Acetylene (ethyne) is the simplest alkyne, and because of its triple carbon–carbon bond is highly unsaturated. You will use the commercial synthesis of this very important organic raw material; the synthesis involves treating calcium carbide with water—that is,

$$CaC_2 + 2H_2O \longrightarrow HC{\equiv}CH\uparrow + Ca(OH)_2$$

After preparing three or four test tubes of each of these hydrocarbons, you will run a series of comparison tests with them.

NOTE TO THE STUDENT: Organic and biological chemistry is somewhat different from the general chemistry of the earlier experiments. The sheer numbers and chemical versatilities of its compounds open the study to many innovative procedural options. Though we stick to specific and proven experimental techniques in the following experiments, we do not spell out precisely the equipment and materials needed for each investigation. Instead, the materials and equipment used will depend on the specific wishes and resources of your instructor.

MATERIALS NEEDED

A pneumatic trough, a mortar and pestle, and a generator tube. The generator tube should be a 6–8 inch pyrex test tube taking a No. 4 one-hole rubber stopper. Glass and rubber tubing for delivery tubes and connections, as illustrated in Figure 15.1, are also needed.

All other reagents, reference compounds, and materials are provided in your bench equipment and on the reagent shelf or bench in the lab.

EXPERIMENTAL PROCEDURE

A. Synthesis

1. *Methane:* Set up an apparatus, as illustrated in Figure 15.1, for generating and collecting methane gas. Note that the clamp on the generator tube is

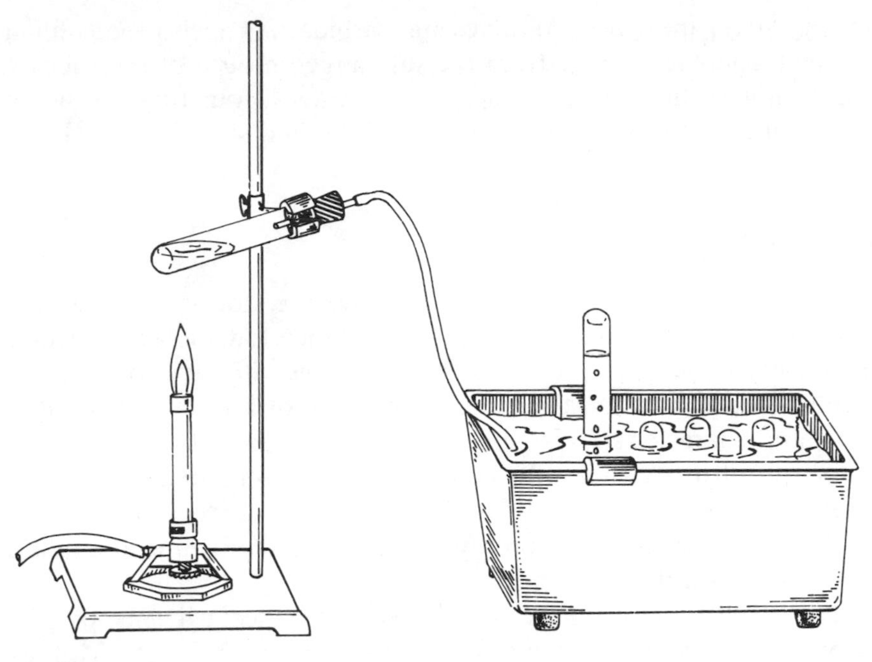

Figure 15.1

fixed as closely as possible to the stopper to minimize the direct heating of the clamp itself. Castaloy clamps, in particular, melt when heated directly with a burner flame. You may also wish to have a pinchcock clamp available for closing the delivery tube after you have collected as much methane as needed. Be certain that all stopper and hose connections are leak free. Be sure that your generator tube is absolutely dry and that no water from the trough siphons or is drawn back into the generator before, during, or after generating the methane. Have rubber stoppers available for the collection tubes. Prepare six water-filled collecting tubes and have them, inverted and in readiness, in the pneumatic trough.

On a top-loader or open-beam balance, weigh out *about* 3 grams of anhydrous sodium acetate and mix this with an estimated equal amount of soda lime in a dry mortar. Grind and mix thoroughly with a pestle and transfer the mixture into the generator tube. Secure the stopper firmly and begin the heating—gently at first—and then at such a rate that a steady stream of gas bubbles flows into the water in the trough. Collect as many tubes of the gas as you can during one continuous heating, letting each filled tube stand in the trough until no more gas is needed. *Without discontinuing the heating,* pinch off the delivery tube and disconnect the stopper from the generator tube. Then, *and only then,* discontinue the heating. Stopper each collection tube, under water, and remove for use in part B. Remember which tube was filled first. This will contain air from the purging of the generator and should be discarded.

2. *Acetylene:* Fill several test tubes with water and invert them in a beaker of water such that no air remains in the tubes. Pipet or siphon away water from the beaker until no more than an inch of water remains. Drop a pea-sized chunk of CaC_2 into the water in the beaker and *quickly* maneuver one of the tubes—without raising its mouth above the water's surface—over the elusive effervescing carbide. As each tube is filled with acetylene (bubbles escape from the submerged mouth of the tube) replace it with another tube. In this way prepare at least four tubes of acetylene gas. Stopper each tube quickly and securely, and save for part B.

B. Comparison Tests

1. *Combustion:* **WEAR GLASSES.** Remove the stopper from a tube of methane and quickly ignite with a match. Since low-molecular-weight alkanes burn with a very quiet blue flame, you will have to observe closely to detect the action. Do the same with a tube of acetylene. The difference in burning characteristics is generally true of saturated as opposed to unsaturated hydrocarbons. If the tubes pop when ignited, you had air mixed with the gas.

   Describe your observations and write complete balanced equations for the reactions on the data sheet.

   Taking care to extinguish nearby burners, place a *few drops* of cyclohexane in an evaporating dish and ignite. Do the same with a *few drops* of toluene and see if the flame characteristics of saturated and unsaturated compounds hold true with liquids as they did for the two gases tested. Write the equations for these combustions also.

2. *Baeyer's test* (mild oxidation): The double or triple bonding of unsaturated compounds constitutes a reactive site. Chemical reagents are more likely to attack such bonds than the less strained carbon–carbon single bonds. Potassium permanganate is a strong oxidizing agent when in the presence of acids or bases. When cold, dilute, and neutral, it is a mild oxidant and is called Baeyer's reagent. As such it will still react with carbon–carbon double and triple bonds, and the sharp color change provides unmistakable evidence for the chemical reaction progress. For example:

$$\underset{\text{An alkene}}{3RCH = CH_2} + \underset{\text{Purple}}{2MnO_4^-} + 4H_2O \longrightarrow \underset{\text{A glycol}}{3\underset{\text{OH}}{RCH} - \underset{\text{OH}}{CH_2}} + \underset{\text{Brown}}{2MnO_2\downarrow} + 2OH^-$$

   Since glycols are also easily oxidized, the reaction often proceeds further. However, the color change is the same for the oxidation of glycols with permanganate, so the test is still valid when applied to the hydrocarbons.

   Add about 0.5 mL of Baeyer's reagent to one of the tubes of methane; quickly stopper with a *clean* cork and shake. Do the same to a tube of acetylene. In your data sheet, describe your observations and write equations for those reaction(s) that took place.

   Try the Baeyer's test on gasoline and on kerosene and record your conclusions. Remember to avoid open flames!

3. *The bromine water test* (a 1,2-addition reaction): When bromine is freshly dissolved in water, a red-brown solution results. As a reactive halogen, the bromine attacks unsaturated sites and saturates them with bromine. In

the process, the color disappears. For example:

$$\text{Cyclohexene (colorless)} + Br_2 \text{ (red-brown)} \xrightarrow[\text{dark or light}]{\text{cold}} \text{1,2-Dibromocyclohexane (colorless)}$$

Saturated compounds react with bromine only under anhydrous conditions and in the presence of sunlight or special catalysts. For example:

$$RH + Br_2 \xrightarrow[\text{ultraviolet light}]{\text{substitution}} RBr + HBr$$

Such substitution reactions do not take place under the conditions of our tests.

By now you have observed that hydrocarbons are not miscible with aqueous reagents. Shaking was necessary to provide enough interfacial surface between the two liquid phases to get a reaction. Another expedient is to dissolve the reagent in a nonaqueous solvent, a solvent that will indeed be miscible with the substance to be tested. Bromine will dissolve in either water or organic solvents. Both reagents are provided, so *run parallel tests* with Br_2/H_2O and with $Br_2/Cl_2CFCClF_2$ (freon 113) and compare their effectiveness as tests for unsaturation. You may observe the demonstration of the extraction phenomenon afforded when benzene is treated with Br_2/H_2O. See Experiment 18 for a discussion of extraction.

Test each of the following for evidence of unsaturation:

CH_4 gas (methane) — Add about 1 mL of bromine water to one of the tubes collected, replace the stopper, and shake. Test a second tube of CH_4 in the same way with bromine in freon 113 ($Br_2/Cl_2CFCClF_2$).

C_2H_2 gas (acetylene) — Test with Br_2/H_2O and $Br_2/Cl_2CFCClF_2$ as was done with methane.

C_6H_6 (benzene) — To 1 mL of Br_2/H_2O add two drops of benzene and mix thoroughly. A good way to do this mixing is to hold the test tube firmly at the rim with the thumb and fingertips of one hand while slapping the bottom of the tube gently with the forefinger of the other hand. This initiates a swirling and mixing motion within the tube. You could also stopper the tube and shake to mix, but as a rule, stoppers introduce impurities and often in sufficient amounts to invalidate tests. Test in the same way with $Br_2/Cl_2CFCClF_2$. (Benzene is highly unsaturated but because of its complete conjugation and resonance is unusually stable.)

C_6H_{12}(cyclohexane)	Try both test reagents on this as you did for benzene.
$CH_3CH = CHCO_2H$ (crotonic acid)	This is a solid compound. Simply add a few crystals to the reagent and mix as before. Test with both Br_2/H_2O and with $Br_2/Cl_2CFCClF_2$.
Gasoline	Test with both reagents.
Kerosene	Test with both reagents.
Safflower oil	Test with both reagents.

You are encouraged to investigate other organic materials with these tests if time and inclination permit. For example, you might check out natural gas, butter, lube oil, turpentine, or your favorite hair oil. Many of these special materials will be available for testing in the laboratory.

DATA SHEET

Name: ______________________ Lab Section: ______ Date: __________

THE PREPARATION AND COMPARISON OF METHANE AND ACETYLENE

Test Reactions

	Combustion	Baeyer's Test	Br_2/H_2O	$Br_2/Cl_2CFCClF_2$
Methane, CH_4	______	______	______	______
Acetylene, $HC{\equiv}CH$	______	______	______	______
Cyclohexane, C_6H_{12}	______	XX	______	______
Benzene, C_6H_6	______	XX	______	______
Gasoline (a mixture)	XX	______	______	______
Kerosene (a mixture)	XX	______	______	______
Crotonic acid, $CH_3CH{=}CHCO_2H$	XX	XX	______	______
Safflower oil (a glyceride)	XX	XX	______	______

Equations: Please complete and balance. If no reaction was observed, indicate by NR.

$CH_4 + O_2 \xrightarrow{\text{complete combustion}}$

$HC{\equiv}CH + O_2 \xrightarrow{\text{complete combustion}}$

(toluene, $C_6H_5CH_3$) $+ O_2 \xrightarrow{\text{complete combustion}}$

$C_6H_{12} + O_2 \xrightarrow{\text{complete combustion}}$

$HC{\equiv}CH + MnO_4^- \xrightarrow{\text{Baeyer's test}}$

$CH_3CH{=}CHCO_2H + MnO_4^- \xrightarrow{\text{Baeyer's test}}$

$CH_4 + Br_2/H_2O \longrightarrow$

$HC{\equiv}CH + Br_2/Cl_2CFCClF_2 \longrightarrow$

$CH_3CH{=}CHCO_2H + Br_2/H_2O \longrightarrow$

Summary statement of test observations (What does each test tell the chemist?):

QUESTIONS

1. A certain liquid acid was found to have a boiling point of 141.3°C. A neutralization equivalent was taken and its molecular weight was established as 73 ±/− 1. The two most probable acids for these findings are:

propionic acid, $CH_3CH_2CO_2H$, bp = 141°, *MW* = 74

and

acrylic acid, $CH_2{=}CHCO_2H$, bp = 141.7°, *MW* = 72

On the basis of what you have learned in this experiment, how might you identify the acid?

2. Ethylene dibromide (1,2-dibromoethane) was synthesized in the laboratory by adding Br_2/H_2O to 3360 mL (STP) of ethylene gas in small portions and shaking until a pale brown color of unreacted bromine persisted:

$$\underset{\text{Ethylene}}{CH_2{=}CH_2} + Br_2 \longrightarrow \underset{\text{1,2-Dibromoethane}}{\overset{\text{Br}\quad\ \ \text{Br}}{CH_2{-}CH_2}}$$

a. How many moles of 1,2-dibromoethane should theoretically form?

b. How many grams of 1,2-dibromethane could theoretically be expected from this synthesis?

c. Pure 1,2-dibromoethane has a density of 2.056 g/mL and a boiling point of 108°. As a typical organic compound it is insoluble in water.

The bottom layer of the liquid in the reaction flask was separated, dried, and distilled to yield 112 mL of a clear, water-white liquid boiling at 108°. What was the actual yield of 1,2-dibromoethane expressed in grams?

d. What was the percent yield in this synthesis?

EXPERIMENT 16

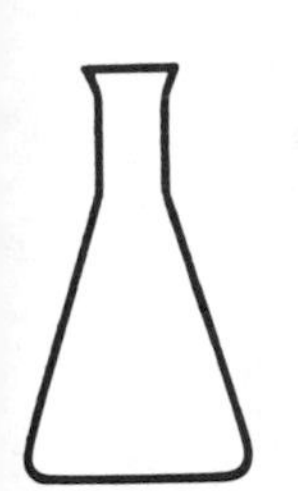

The Hydrocarbons and Their Halogen Derivatives: The Identification of an Unknown

OBJECTIVES

1. To gain practice in determining and utilizing fundamental physical properties for the identification of organic compounds.
2. To acquire an appreciation of the secondary reference sources as aids in chemical work.
3. To learn how to devise a plan for the possible identification of a hydrocarbon or a halogen compound.

PRELABORATORY QUESTIONS

1. Define boiling point. Freezing point. Melting point.
2. List two tests by which you might establish whether a certain liquid is saturated or unsaturated.
3. How does the presence of halogen in an organic compound affect the density of that compound?
4. How might you distinguish a saturated alkane from an aromatic hydrocarbon?

DISCUSSION

You have learned that organic compounds fall into distinct classes based on functional groups. Each chemical class has its own spectrum of chemical reactions. The chemist uses a series of classification tests to establish the class to which a chemical must belong. Once this is done, determination of its unique physical and chemical properties will complete its identification. This is the essence of qualitative organic analysis.

In this experiment you will carry out a few well-known classification tests and some techniques for measuring physical properties. With these tests and a mind for selective elimination, this experiment should be an interesting challenge.

You will be given a liquid from the list of compounds in Table 16.1. Identify it by any means at your disposal. It is suggested that you first establish its boiling point and density and then apply the process of elimination. The presence of unsaturated sites, or of halogen, can usually narrow the choice to one or two possibilities. When you are quite sure of your identification, look up your compound (or likely compounds) in a handbook or in some other library reference such as *Heilbron's Dictionary of Organic Compounds* or *The Merck Index*. Ask your laboratory instructor for assistance if needed. The additional information given often suggests further simple qualitative tests or provides data to verify your identification.

TABLE 16.1 List of Possible Unknowns (Compound, Boiling Point, and Density)

Alkanes: C_nH_{2n+2}; *RH*	T_b (°C)	*d (g/mL)*
n-pentane	36.2	0.626
n-hexane	69	0.660
n-heptane	98.5	0.684
n-octane	125.9	0.704
n-nonane	150.7	0.718
n-decane	174	0.730
Cycloalkanes: C_nH_{2n}		
cyclopentane	50	0.745
cyclohexane	81.4	0.779
methylcyclohexane	100.3	0.786
Cycloalkenes: C_nH_{2n-2}		
cyclopentene	45	0.774
cyclohexene	83.3	0.810
4-methylcyclohexene	102	0.801
Aromatic hydrocarbons: ArH		
benzene	80	0.879
toluene	110.8	0.866
ethylbenzene	136	0.867
p-xylene	138	0.861
o-xylene	144	0.875
mesitylene	165	0.865

Alkenes: C_nH_{2n}	T_b (°C)	*d (g/mL)*
1-pentene	30	0.645
1-hexene	64.1	0.673
1-heptene	95	0.699
1-octene	126	0.722
1-nonene	149.9	0.730
1-decene	172	0.763
Halogen compounds: RX, ArX, etc.		
ethyl bromide	38	1.43
n-propyl chloride	47.2	0.890
t-butyl chloride	51	0.847
chloroform	61.3	1.50
n-propyl bromide	70.9	1.353
n-butyl chloride	78	0.884
carbon tetrachloride	76	1.60
ethylene chloride	83.5	1.257
n-butyl bromide	101.6	1.299
n-butyl iodide	131	1.617
ethylene bromide	131.6	2.17
chlorobenzene	132.1	1.107
bromoform	149.5	2.89
carbon tetrabromide	189.5	3.42

MATERIALS NEEDED

Check out a vial of an unknown organic hydrocarbon or halide. All necessary reference compounds, reference handbooks, reagents, and instruments are available in the laboratory, and equipment not included in your lab locker can be checked out as needed.

EXPERIMENTAL PROCEDURE

Many of these liquids are very flammable! Note the location of the fire extinguishers and deluge showers before beginning this experiment. Keep the bulk of your sample covered and away from the flames. However, since small quantities are always specified, possible fires are easily blown out after removing their cause. Avoid breathing vapors and allowing skin contact with *all* chemicals.

Try some of the tests described below on known reference compounds first, and write the equations as a means of study. These need not be turned in with your report, but are meant to point out the more important observations and principles behind this experiment. If you need help, consult your laboratory instructor. Ideally, you should let your lab instructor correct any errors you might have made.

A. Boiling Point

In general, the lower the molecular weight of a compound, the lower its boiling point will be. Thus, in most homologous series the boiling point is raised about 25° to 30° C for each additional CH_2 group. Inspection of the table of unknowns will verify this rule of thumb.

An accurate boiling point is best determined with a distillation apparatus. In this experiment sufficient accuracy can be attained as follows.

Clamp a *tall*, clean, *dry* test tube in position over a small hole in a wire gauze, as illustrated in Figure 16.1. Place about 2 mL of your unknown in the tube. The thermometer is fixed in position so that the bulb does not touch the sides of the test tube, and is about 1 inch from the surface of the liquid contents of the tube. Place a boiling chip in the tube.

Carefully, with a low flame, heat the test tube by playing the flame directly on the hole in the gauze. The heating should be steady until the liquid boils gently. Observe the rise of the condensing vapor and continue a slow, steady heating until the mercury bulb of the thermometer is completely bathed in the condensing vapor and the temperature becomes constant. Carefully regulate the heating such that the condensation level is only a few centimeters above the mercury bulb of the thermometer. Record your temperature on the data sheet as soon as it has leveled to a constant, and quickly discontinue heating. This is the *boiling point, that temperature at which the vapor pressure of the liquid equals the atmospheric pressure*. You can use the boiled sample in the tube for subsequent tests.

Boiling points vary with changes in barometric pressure and vary slightly depending on the purity of the compound and the method of determination. Thus your experimental result could deviate as much as 5° from the literature

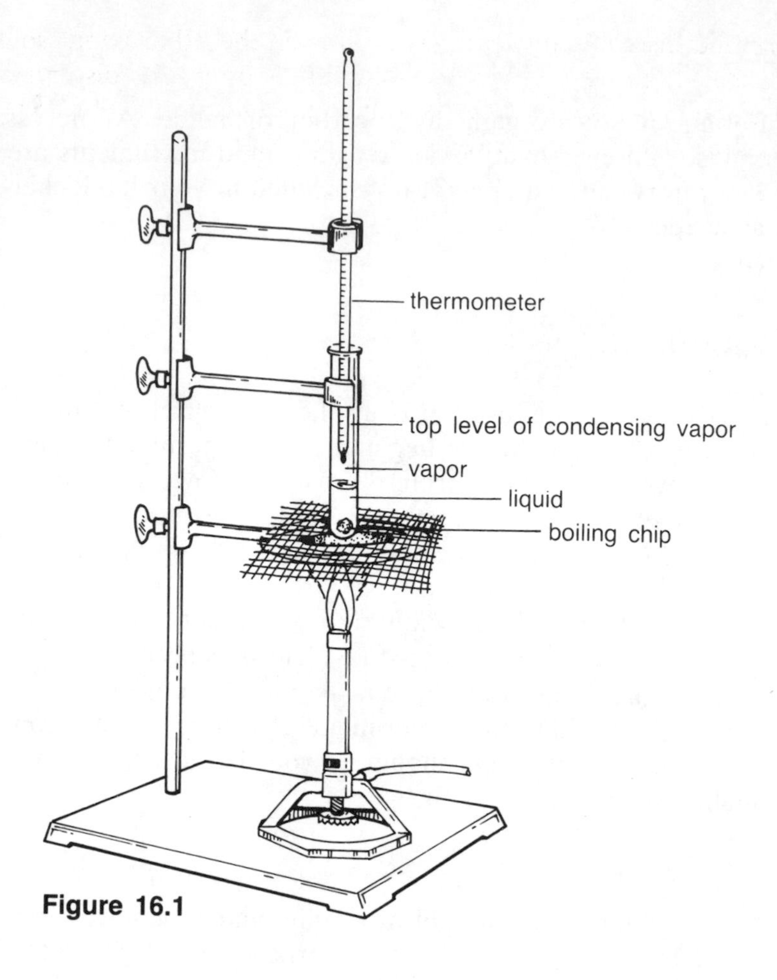

Figure 16.1

value. You will do well to record the most probable boiling point *range* and use this to predict likely possibilities.

B. Density

The density of a liquid may be determined with accuracy using the methods described in Experiment 2. If you wish, you may make or check out a pycnometer and establish a more reliable density. However, many of the densities listed for the compounds are so close to one another that valid conclusions as to identity are difficult even when accurately determined. An inspection of the table does disclose one very useful generalization. *All hydrocarbons have densities less than unity*. Except for a few of the monochlorides, the halogen derivatives have densities greater than 1. If, then, your compound floats on water (and all of the unknowns are insoluble in water), in all probability it is a hydrocarbon or one of the monochlorides. If it sinks, it must be one of the other halogen compounds. Simply add a couple of drops of unknown to a couple of milliliters of water in a test tube, tap the tube, and observe. Record your observations on the data sheet.

C. Solubility

In general, the hydrocarbons and the halogen derivatives are insufficiently polar to permit water solubility. All compounds used in this experiment are typically organic and insoluble in water. They are, on the other hand, soluble in organic solvents. "Like dissolves like." Solubility is useful for distinguishing alkanes and alkyl halides from *other* classes of compounds but is not helpful for this experiment. It is included here for your information only.

D. Tests for Unsaturation

Alkenes and alkynes are relatively reactive at the site of their double and triple bonds. They react by "addition" to form a more stable, saturated derivative that allows a tetrahedral bond arrangement about the carbon atoms involved.

Aromatic hydrocarbons, even though highly unsaturated, are so stabilized by resonance that they behave chemically as though they were saturated.

Baeyer's Test for Unsaturation Unsaturated compounds are readily oxidized to glycols (1,2-dihydroxy compounds) by cold dilute neutral permanganate solution. Baeyer's reagent is simply a 0.1% aqueous solution of $KMnO_4$. In general, the equation for the oxidation of the unsaturated compound is as follows:

$$3R_2C{=}CHR + 2MnO_4^- + 4H_2O \longrightarrow 3R_2\underset{\displaystyle OH}{\underset{|}{C}}{-}\underset{\displaystyle OH}{\underset{|}{C}}HR + 2MnO_2\downarrow + 2OH^-$$

To test an unknown liquid for unsaturation, add one or two drops to 1 mL of Baeyer's reagent in a test tube and shake. If a carbon-to-carbon double or triple bond is present, the purple of the MnO_4^- fades and a brown precipitate of MnO_2 forms. Try this test on your unknown and on an appropriate unsaturated compound from the reference shelf. Record your observations on the data sheet.

The Decolorization of Bromine Water Unsaturated compounds react by addition with Br_2, and the progress of the reaction can be followed by the fading of the red-brown bromine color. It should be noted that Br_2 is more soluble in organic solvents than it is in water. The red-brown color, therefore, first transfers into the organic layer (a good illustration of the process of *extraction*) and then fades as the bromine adds to the site of unsaturation. A generalized test reaction might be written as follows:

$$R_2C{=}CR_2 + Br_2 \longrightarrow R_2\underset{\displaystyle Br}{\underset{|}{C}}{-}\underset{\displaystyle Br}{\underset{|}{C}}R_2$$

Sometimes the bromine is dissolved in freon 113 instead of water. This modification is especially useful when one wishes to test a solid organic compound for unsaturation. Why? If the answer is not apparent to you, try both bromine in freon 113 and Br_2/H_2O on small samples of cinnamic acid, $ØCH{=}CHCO_2H$ (the symbol Ø is used here to abbreviate $C_6H_5^-$), and see for yourself which is best for the detection of the unsaturation. Be sure tubes and stoppers are clean! Add a couple of drops of the compound to be tested to 1 mL of the bromine solution and shake.

Ignition Test This is, at best, a very qualitative and uncertain test. However, it is a fact that the more unsaturated a compound is, the more incandescent and smoky the flame when the compound is burned in air. Alkanes generally burn quietly with a blue to blue-yellow flame when ignited in an evaporating dish. Aromatic compounds burn with a *smoky* yellow flame. For example, kerosene for use in lamps is a mixture of C_{12} to C_{18} hydrocarbons. If kerosene is formulated with too few alkene components, its flame does not illuminate well because the number of incandescent particles in the flame are too few. If the alkene and/or aromatic fraction is too large, the flames are very smoky and blacken the chimneys with soot.

Observe your burner flame. Natural gas is about 80% methane with smaller amounts of ethane, propane, and butanes.

With this test you can distinguish an alkane from an aromatic hydrocarbon. Try it, but run some controls on known samples first, and *use small quantities*—unless you enjoy cleaning up soot! Use a match to ignite about half a milliliter of the compound to be tested contained in an evaporation dish. Observe and record your observation.

E. Tests for Halogen

The presence of halogen in an organic compound can be detected by subjecting the compound to conditions rigorous enough to release the halogen.

The Beilstein Flame Test A clean copper wire is inserted into a cork to provide an insulated handle. The wire is then heated to red heat in a burner flame and quickly, while still red hot, dipped into a small amount of the compound to be tested for halogen. The hot copper reacts with a little of the organic halide to produce a thin film of cuprous halide on the surface of the wire. Cuprous halides impart a blue-green color to a flame. Thus, if halogen is indeed present, when one reheats the wire in the oxidizing flame of the burner, a fleeting blue-green flame is observed. This test is not unequivocal, however, and the presence of halogen, if indicated, should be verified by other tests. Certain other classes of compounds, such as the amines, can also produce a blue to green flame in the presence of copper. Try this test on your unknown and on an appropriate halogen-containing reference compound.

The Alcoholic Silver Nitrate Test Organic halides react at different rates with silver nitrate to form silver halide precipitates. The general reaction for this test is:

$$RX + Ag^+ \longrightarrow AgX\downarrow + \text{organic by-products}$$

Alcoholic silver nitrate is preferred because organic compounds are typically more soluble in alcohol than they are in water, and the reactions can take place more rapidly. In general, iodides react readily, bromides less readily, and chlorides with difficulty. Aromatic halides scarcely react at all, even under reflux.

To a milliliter of alcoholic silver nitrate in a test tube, add two drops (or a couple of crystals) of the compound you wish to test. Tap the tube to facilitate mixing and observe closely. A white precipitate (which may form quite slowly) indicates that chlorine is present. A canary yellow precipitate results when iodides are tested. Bromides form a very pale yellow precipitate that is easily mistaken for silver chloride. If no precipitate develops, heat the mixture gently for a few minutes. A faint cloudiness that develops immediately is most probably the result of a dirty test tube. A faint cloudiness that develops with time suggests the presence of one of the less reactive halides.

Oxidation to Free Halogen An oxidizing agent powerful enough to decompose organic compounds will, in the process, release any halogen present as free halogen:

$$2C_3H_7X + 19MnO_2 + 38H^+ \xrightarrow{\text{heat}} 6CO_2\uparrow + 19Mn^{2+} + 26H_2O + X_2\uparrow$$

To run this test, place approximately 0.1 g of powdered MnO_2 in a *clean, dry* test tube and add to this a few drops (or crystals) of the compound to be tested. Mix thoroughly with a glass stirring rod. To the mixed paste or powder, next add about 1 mL of concentrated H_2SO_4 and mix again. If a reaction starts, as evidenced by a significant rise in temperature, watch for the appearance of vapors in the tube. Violet vapors are I_2, red-brown vapors are Br_2, and pale green vapors are Cl_2. The color of the Cl_2 vapors is often undetectable, but the odor is quite unmistakable. If nothing significant is observed, heat the mixture gently in a flame but be careful to avoid spattering. **(Wear safety glasses! Be cautious when testing odors!)**

F. Refractive Indices

If a refractometer is provided, enlist your instructor's help in determining the refractive index, n, of your unknown compound. Handbooks and other secondary reference sources list refractive indices for liquids, and these are often valuable diagnostic aids in identification. The refractive index is an easily measured physical property of a compound, and although the range of refractive indices is small in view of the enormous number of organic liquids, marked and predictable differences often distinguish between compounds in a small select list.

When finished with the refractometer, be sure to rinse the prism and the instrument window thoroughly with methanol and blot dry with lens paper (do not scrub!). Organic liquids soften the cement used to fix the instrument's prisms and lenses and can do irreparable damage if allowed enough time to act on cements or finishes.

The refractive index of a liquid is a function of the degree to which a beam of incident light is bent (or refracted) as it passes from air into that liquid. The refractometer enables one to read this refractive index directly on a dial.

DATA SHEET

Name: ______________________ Lab Section: ____ Date: __________

THE HYDROCARBONS AND THEIR HALOGEN DERIVATIVES: THE IDENTIFICATION OF AN UNKNOWN

Unknown Number	Name of Unknown	Formula of Unknown
________	________	________

Test	Observed Result	Literature Value or Prediction
Boiling point, T_b	________	________
Density, d	________	________
Baeyer's test	________	________
Br_2/H_2O test	________	________
$Br_2/Cl_2CFCClF_2$ test	________	________
Ignition test	________	________
Beilstein test	________	________
Alcoholic $AgNO_3$ test	________	________
Oxidation to free halogen	________	________
Refractive index, n	________	________

Other observations pertinent to the identification; tests and uncertainties:

Equations for chemical tests that were positive:

QUESTIONS

1. A compound listed in Table 16.1 was found to be heavier than water. Its boiling point approximated 60° C, and it tested negatively with Baeyer's reagent and with bromine water. A white precipitate slowly developed when it was heated with alcoholic silver nitrate. Which compound must it be?

2. Another compound from the list in Table 16.1 is lighter than water and gives a negative Beilstein test as well as a negative test with alcoholic silver nitrate. When shaken with cold, dilute, and neutral potassium permanganate, a dark brown precipitate forms. It boils at approximately 150° C. What is its identity?

3. Yet another compound from Table 16.1 gives negative tests with bromine in freon 113, with alcoholic silver nitrate, and with Baeyer's reagent. It is lighter than water, forms a lot of soot when it burns in an open flame, and when distilled comes over in the range 77–81° C. Which compound must it be?

4. Another compound from Table 16.1 was found to boil at about 50° C by the crude technique used in this experiment. Its refractive index was determined and found to be 1.3887. Which compound best accommodates these data? (Look up n_D in a handbook.)

5. The data listed for question 4 are very limited. What might you do in the lab to confirm (or compromise) the tentative identification made for question 4?

6. Suppose you are investigating an unknown white solid. How can you establish whether it is organic or inorganic in nature?

The Iodine Number of a Lipid: Wijs Method

OBJECTIVES

1. To illustrate how unsaturation is quantitated for lipids.
2. To review organic addition reactions.
3. To review oxidation-reduction and its stoichiometry.
4. To gain familiarity with lipids as a class of organic compounds.

PRELABORATORY QUESTIONS

1. Write the balanced equation for the saturation of isoprene (2-methyl-1,3-butadiene) with iodine (I_2).
2. Write the balanced equation for the saturation of isoprene with Wijs reagent (ICl).
3. Write a balanced equation for the reduction of I_2 with thiosulfate ($S_2O_3^{2-}$).
4. What should be done if you get Wijs reagent on your hands?

DISCUSSION

The fats, oils, and waxes are all esters of plant or animal origin. As representative lipids they are water insoluble and therefore chemically unique compared to carbohydrates, proteins, and other chemicals endemic to living organisms. Table 17.1 lists 20 of the most common triglycerides. An inspection of this table will show that if two measurable chemical properties of a triglyceride (its iodine number and its saponification number) are known, its identification is essentially established. In this experiment you will determine the iodine number for an unknown triglyceride. You will be given the melting point range and the saponification number for your unknown. With this given information and your laboratory results, you can identify your

TABLE 17.1

Fat or Oil	Saponification #	Iodine #	Melting Point Range
Lanolin	82–130	17–29	41–
Cod liver oil	171–189	137–166	21–38
Castor oil	175–183	84	13
Olive oil	185–196	79–88	26–30
Peanut oil	186–194	88–98	–5–
Corn oil	187–193	111–128	17–20
Linseed oil	188–195	175–202	20–24
Soy bean oil	189–193	122–134	26–28
Tung oil	190–197	163–171	40–44
Cocoa butter	193–195	33–42	48–53
Lard	193–198	63–79	33–38
Neat's-foot oil	193–199	58–75	29–41
Cottonseed oil	194–196	103–111	34.5
Mutton tallow	195–196	48–61	34–49
Beef tallow	196–200	35–42	43–44
Chaulmoogra oil	196–213	98–110	–25
Palm oil	200–205	49–59	50
Butter fat	210–230	26–28	38–41
Coconut oil	253–262	6–10	24–27
Tributyrin (pure)	549	0	–75

triglyceride as one of those listed in Table 17.1. Later, in Experiment 23, the saponification process is studied separately.

By definition, the *iodine number* of a substance is the number of grams of iodine needed to saturate 100 g of that substance. In other words, the iodine number (sometimes called iodine value) is the weight percent of iodine absorbable by the substance. It is a measure of the unsaturation of the substance. The *saponification number*, on the other hand, is a measure of the molecular weight of the substance and is defined as the number of milligrams of KOH needed to complete the saponification of 1 gram of an ester (fat, oil, or wax).

Naturally occurring lipids are seldom pure compounds. Rather, they are mixtures of many related compounds. As a consequence, their measured physical properties are statistical averages, and their chemical properties are reported as ranges. A look at Table 17.1 confirms this. The listed melting point ranges are just what one might predict for chemical mixtures. The saponification numbers and the iodine numbers are also given as ranges since a specific oil may vary in composition depending on its origin. However, careful measurements of these three properties serve to, at least tentatively, identify a naturally occurring fat or oil. Remember that the fats are triglycerides that are relatively saturated and are therefore solids at room temperature. Oils, on the other hand, are more unsaturated and exist as liquids at room temperature.

The Wijs method for determining iodine number will be used. In this procedure, Wijs reagent (ICl) is substituted for the slower acting I_2, but results are calculated as though I_2 had been used to saturate all double bonds instead of the ICl actually employed. **Should you get any of this reagent on your skin, wash first with the dilute sodium thiosulfate solution provided for that purpose and then rinse generously with water**.

Recall the bromine tests for unsaturation in Experiment 15. The Wijs method applied in this experiment is a refinement of this test for unsaturation—a quantitative application of the same addition reaction. In the Wijs method, ICl is a compromise reagent. Bromine is too reactive for lipids and cannot be prevented from engaging in some substitution reactions on the side. Iodine is too sluggish and does not react quantitatively. Iodine monochloride is between Br_2 and I_2—moderately fast and well behaved. In the dark it confines its activities to addition reactions.

$$\underset{x \text{ moles}}{\text{—CH=CHR}} + \underset{\text{excess}}{\text{ICl}} \longrightarrow \underset{x \text{ moles}}{\text{—CH(I)—CHR(Cl)}} + \underset{\text{unreacted excess}}{\text{ICl}} \qquad (1)$$

Iodine monochloride, then, acts as a stand-in for iodine. Actually, in our titrations with thiosulfate, ICl behaves as though it were actually I_2. Consider the reactions that take place when x moles of ICl are titrated, as in this procedure. First, an excess of KI is added. The ICl is quantitatively converted to I_2:

$$\underset{\text{moles}}{\text{ICl}} + \underset{\text{excess}}{\text{KI}} \longrightarrow \underset{x \text{ moles}}{I_2} + \underset{x \text{ moles}}{\text{KCl}} + \underset{\text{leftover excess}}{\text{KI}} \qquad (2)$$

The I_2 (the same number of moles as there were moles of ICl at the start) is then titrated with the standard thiosulfate to the starch end point:

$$\underset{\text{blue}}{I_2 + 2Na_2S_2O_3} \xrightarrow[\text{indicator}]{\text{starch}} \underset{\text{colorless}}{2NaI + Na_2S_4O_6} \qquad (3)$$

A little reflection should convince you that it is legitimate to pretend, as we do, that the ICl reagent is actually I_2. Both titrate identically in the Wijs method. That is, mole for mole they have exactly equal abilities to oxidize thiosulfate.

In practice, then, there are three operations. First, a fixed volume of ICl reagent is allowed to react in the dark with a known excess weight of unsaturated material until saturation is complete. Second, an equal volume of ICl "blank" is given the same treatment. Third, both runs are titrated with standard thiosulfate solution to establish the final ICl (I_2) contents. The difference between the two determinations is a measure of the ICl absorbed by the sample.

The calculations for the iodine number then follow the usual rules of volumetric analysis. Remember that: mL $\times N =$ meq; meq/1000 = equivalents, and the equivalent weight of I_2 is 253.8/2, or 126.9 g per equivalent.

To calculate the grams of I_2 absorbed, then, since mL $\times N_{thio}$ = meq of thio = meq of I_2, we can make the proper substitutions and get:

$$\frac{(\text{mL}_{\text{thio for blank}} - \text{mL}_{\text{thio for reaction}}) \times N_{\text{thio}} \times 126.9}{1000} = g_{I_2} \text{ absorbed}$$

And since we want the g_{I_2} needed for 100 g of sample (iodine number),

$$\frac{g_{I_2} \text{ absorbed}}{\text{iodine number}} = \frac{g_{\text{sample}}}{100}$$

Solving for iodine number, substituting terms, and canceling, we then arrive at the expression for making our calculation, namely:

$$\text{iodine number} = \frac{(\text{mL}_{\text{thio for blank}} - \text{mL}_{\text{thio for Rx}}) \times N_{\text{thio}} \times 12.69}{g_{\text{sample}}}$$

MATERIALS NEEDED

Pick up a buret, buret holder, a 25 mL pipet, two 500 mL Erlenmeyer flasks, and a vial of unknown lipid. Standard sodium thiosulfate solution, 1 N KI, Wijs reagent, and starch indicator solution are provided in the lab.

EXPERIMENTAL PROCEDURE

The procedure requires the use of C Cl_4 as the reaction solvent, so be very careful to avoid skin contact! There are three single steps.

A. Standardization of Wijs Reagent (Blank Run)

Your instructor may have run the blank for you in order to expedite your work. In that event, record his or her posted volume of thiosulfate needed for a 25.0 mL aliquot of Wijs reagent, and skip to part B. If this is *not* the case, proceed as follows:

In a clean, dry 500 mL Erlenmeyer flask, place 20 mL of CCl_4 and exactly 25.0 mL of Wijs reagent. These materials will be available in automatic dispensing burets. Mix, cover with a clean, dry, overturned beaker, wrap in a towel, and let stand for 30 minutes in a dark place (closed locker).

While this is standing, set up the buret as shown in Figure 17.1 (only one buret is needed). Fill the buret with standard thiosulfate after rinsing first with distilled water and then with a small amount of the standard thiosulfate solution. While you wait, you can also weigh the *clean, dry* sample vial of unknown.

When the blank has stood for 30 minutes, retrieve it and add 20 mL of 1 N KI and 100 mL of distilled water. Mix by swirling. The solution should have a brown color at this point due to the unreacted I_2. Titrate the blank with standard sodium thiosulfate, mixing well between additions until the yellow of the last of the iodine is *almost* gone. Then add 1 mL of starch indicator solution. It should turn to a deep blue color. Continue the titration,

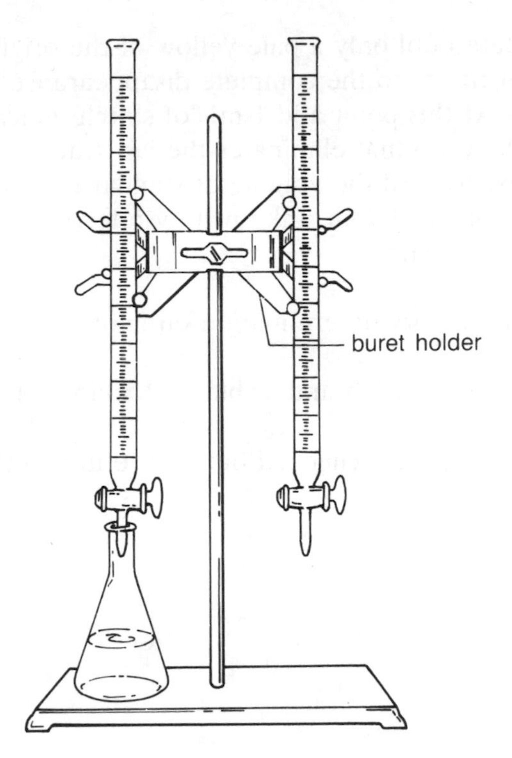

Figure 17.1

drop by drop now and with agitation between additions, until the last of the blue starch–iodine complex disappears.

Record the volume of thiosulfate used and its normality on the data sheet.

B. Saturation of the Sample with ICl

Wipe all traces of oil from the outside of the vial containing your unknown, and then weigh it on the analytical balance. Transfer from this vial an estimated three drops (0.2 mL) of the unknown directly into a clean, dry, 500 mL Erlenmeyer. *Do not use a pipet!* Pour directly into the Erlenmeyer such that *all* of the oil transferred falls to the bottom of the flask and does not stick in the flask's neck or run down the edges of the vial. It is especially important that *all* of the sample removed goes into the Erlenmeyer flask!

If your particular unknown is a solid, you will have to melt it by gently warming in a beaker of heated water before pouring out a sample for analysis. Take care that no water contaminates your unknown or adheres to the outside of the vial when you subsequently make a weighing. Close the vial and weigh it again. The mass of sample taken is, of course, the difference between the two weighings.

To the sample in the Erlenmeyer, add 20 mL of CCl_4, and then exactly 25.0 mL of the Wijs reagent from the dispensing buret. Swirl gently to bring about complete solution of the sample and cover the Erlenmeyer with a clean, dry beaker. Wrap the apparatus with a towel and allow to stand in a dark place (bench drawer) for 30 minutes.

C. Titration of Reaction Mixture with Standard Thiosulfate

To the completed reaction mixture (after 30 minutes) add 20 mL of 1 N KI and 100 mL of distilled water. Mix well and titrate with intermittent mixing with standard thiosulfate until only a pale yellow of the original iodine brown color remains. Do not titrate to the complete disappearance of the yellow, for some I_2 must remain. At this point add 1 mL of starch indicator solution and cautiously titrate to the drop that eliminates the last trace of the blue starch–iodine complex. Record the volume of standard thiosulfate used and calculate the iodine number of your unknown by referring to the equation derived in the discussion section.

Run a second confirmatory determination on another sample as described in parts B and C.

Refer to the discussion section and Table 17.1 to identify your unknown as one of the 20 listed.

Clean and dry all equipment checked out and return to the storeroom.

DATA SHEET

Name: ______________________ Lab Section: ____ Date: __________

THE IODINE NUMBER OF A LIPID: WIJS METHOD

Unknown number: ______________ Unknown is: ______________

$N_{thiosulfate}$ = ______________ eq wt $_{I_2}$ = ______________

	Blank	**Trial 1**	**Trial 2**
Weight of sample			
Volume of Wijs reagent used			
Volume of std thiosulfate used			
meq of I_2 absorbed by sample			
Grams of I_2			
Iodine number			

Miscellaneous observations:

QUESTIONS

1. Complete the equations for (a) the reaction of excess Wijs reagent and (b) the saponification reaction on glyceryl trioleate (olein):

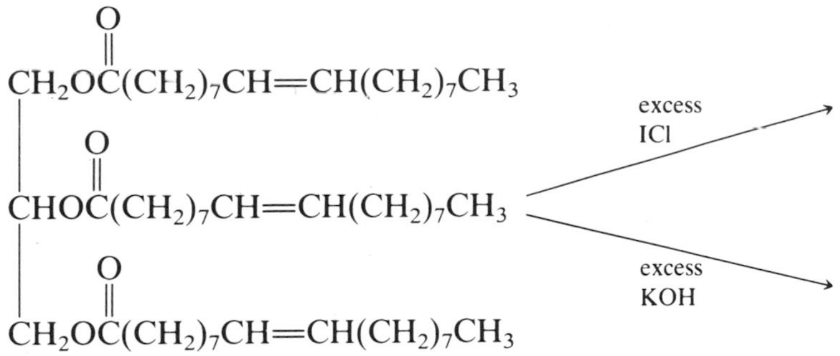

$$\begin{array}{l} \overset{O}{\overset{\|}{CH_2OC(CH_2)_7CH{=}CH(CH_2)_7CH_3}} \\ | \\ \overset{O}{\overset{\|}{CHOC(CH_2)_7CH{=}CH(CH_2)_7CH_3}} \\ | \\ \overset{O}{\overset{\|}{CH_2OC(CH_2)_7CH{=}CH(CH_2)_7CH_3}} \end{array} \quad \begin{array}{l} \xrightarrow{\text{excess ICl}} \\ \\ \xrightarrow{\text{excess KOH}} \end{array}$$

2. Calculate the iodine number for olein (structure in question 1). (MW = 885)

3. Write a balanced equation for the reduction of Br_2 with sodium thiosulfate.

EXPERIMENT 18

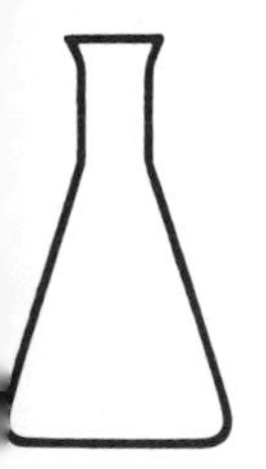

The Separation of a Mixture by Extraction Techniques

OBJECTIVES

1. To review and make practical application of the acid-base interactions of simple organic compounds.
2. To demonstrate the separation and isolation of the acidic, the basic, and the neutral components of a simple mixture.
3. To gain experience in the recovery of solutes from solutions.

PRELABORATORY QUESTIONS

1. Write an equation illustrating the reaction of an organic amine, RNH_2, with dilute sulfuric acid.
2. Write an equation illustrating the reaction of an organic acid, RCO_2H, with potassium hydroxide.
3. Sugar, $C_{12}H_{22}O_{11}$, and salt, NaCl, cannot be separated using the extraction procedure of this experiment. Why not?
4. Sometimes in an extraction process the question arises: Is the water layer on the top or is it on the bottom? Describe how you might unequivocally establish which layer was on the bottom—and in a way that would require no sacrifice of either layer.

DISCUSSION

Nature's liquids are solutions and her solids are usually mixtures. It is no wonder that so much research has gone into the development of separation processes. One of the best of these, and one with limitless versatility, is extraction.

This experiment is the classic technique for the separation of mixtures of acids, bases, and neutral compounds. In the procedure, you alter the solutes to accommodate the solvents, and the separation efficiency becomes almost quantitative.

As a general rule, monofunctional organic compounds (compounds with only one functional group) having more than five carbons tend to be ether soluble and water insoluble. Salts and polyfunctional compounds have the reverse solubility characteristics. They tend to be water soluble and ether insoluble . Herein lies the secret of this widely used extraction technique for the separation of mixtures. The mixture is dissolved in ether (or some other suitable water-immiscible organic solvent). First the amines (organic bases) are converted to water-soluble salts with HCl and are extracted in the water layer. Next the organic acids are neutralized to water-soluble salts with NaOH and extracted in the water layer. Finally, only the neutral compounds remain in the ether and can be recovered by distilling off the ether.

Suppose you had a mixture of an acid (RCO_2H), an amine (RNH_2), and a neutral compound (RH). A flow diagram for its separation should put the procedure into perspective:

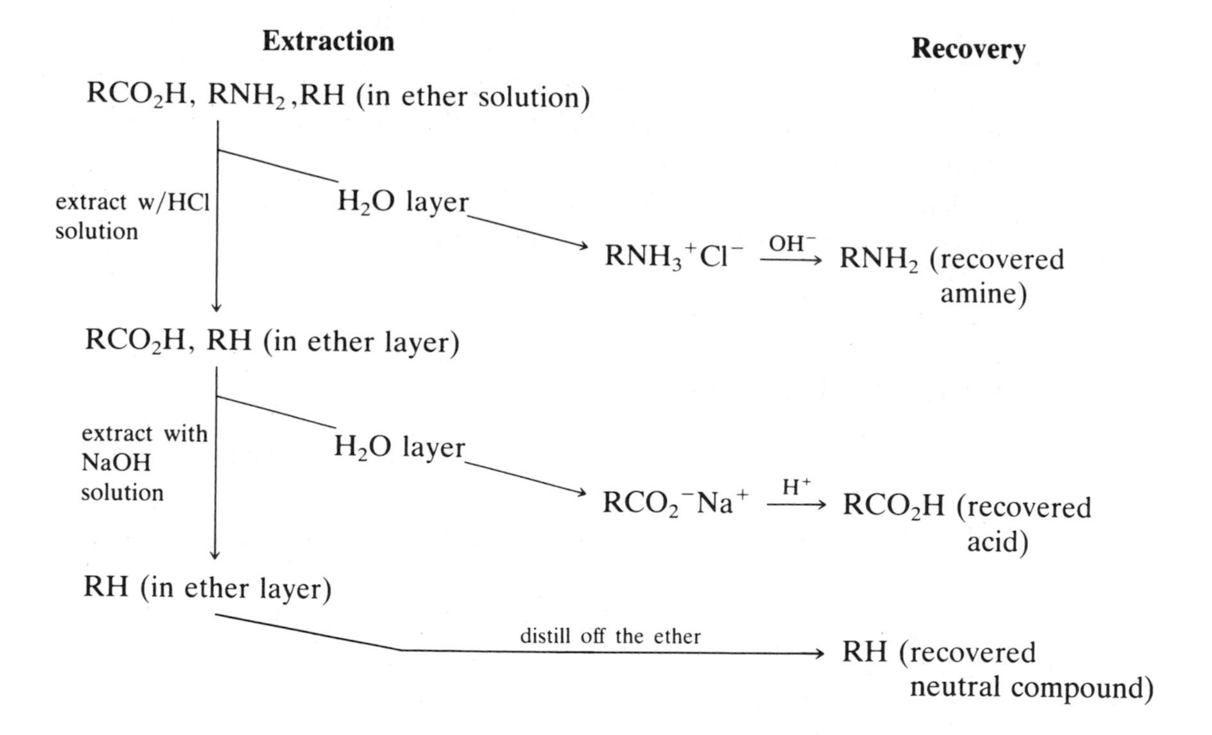

This basic procedure will work on any such mixture as long as the components of the mixture are all ether soluble. Study the flow diagram until you truly understand what is involved. Then, and only then, will the following procedure make sense to you. Then, too, you will understand why you must wash each extraction before recovering the solute.

Unless otherwise informed, you will be issued 4 g of a solid mixture of benzoic acid, *p*-nitroaniline, and naphthalene. For this experiment the mixture will consist of 33.3% by weight of each component, but you are not required to make a quantitative account of your recovery unless you wish to do so for your own satisfaction. Simply separate the mixture into its three component parts. You have enough sample to perform the separation twice on the scale designated.

Be extremely careful to avoid ether fires. Make certain that all fires in the lab are extinguished and that ether bottles and ether solutions are kept cool and covered. Remember, ether boils at 35°.

MATERIALS NEEDED

Obtain at the storeroom 4 g of the mixture of *p*-nitroaniline, benzoic acid, and naphthalene. Also needed will be a 125 mL separatory funnel, a small Büchner or Hirsch funnel, and a filter flask. The ether, 5% HCl, and 5% NaOH solutions as well as other specified materials will be available in the laboratory.

EXPERIMENTAL PROCEDURE

Before beginning this experiment double-check to make certain that *all* flames in the laboratory are extinguished—and that they remain extinguished!

Check the separatory funnel stopcock for proper fit. Lubricate if the stopcock is of glass (but do so very sparingly to avoid plugging the hole with grease). Get some crushed ice in a 600 mL beaker to use as a cooling bath for ether solutions. **Turn off all burners in the vicinity!** Set up the filter flask and the funnel for suction filtration as shown in Figure 18.1.

Dissolve about half (2 g) of the mixture in 50 mL of cold ether. Do not heat, but if necessary to complete the dissolution, add a bit more ether. Cool the solution on an ice bath.

Transfer this cold solution to the separatory funnel set up as in Figure 18.2.

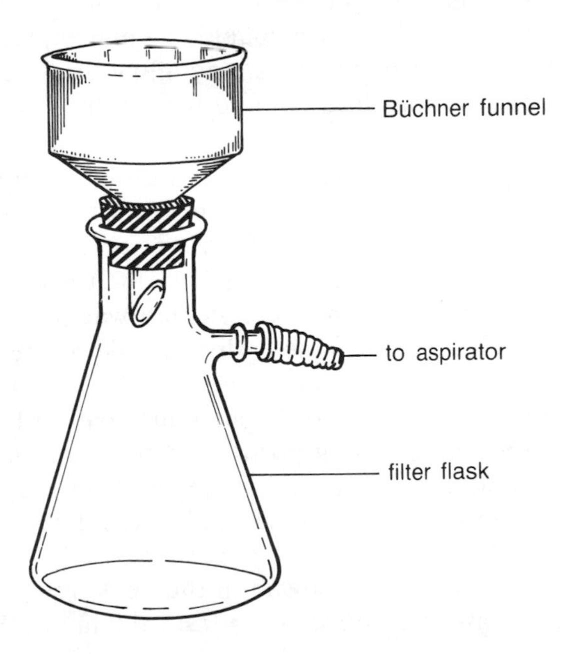

Figure 18.1

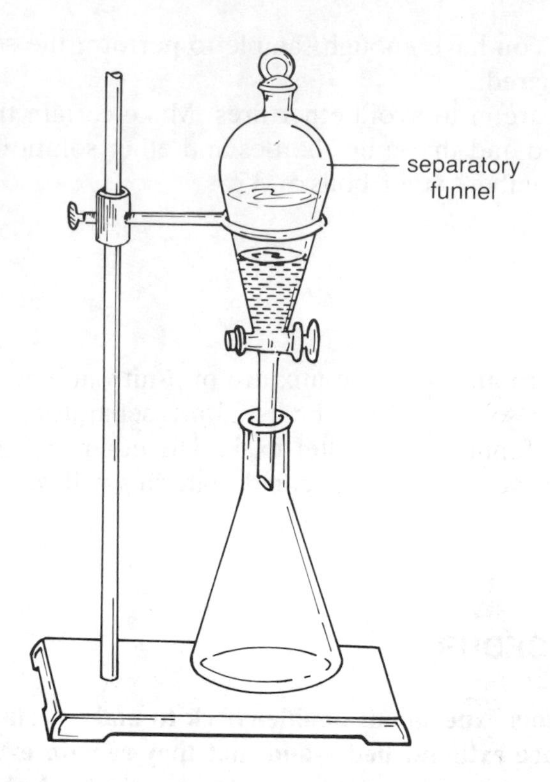

Figure 18.2

A. Isolation of the *p*-Nitroaniline

Add 25 mL of *cold* 5% HCl to the ether solution in the separatory funnel and stopper the top. Using the technique demonstrated by your instructor and illustrated in Figure 18.3, gently shake the separatory funnel, with a swirling motion. Keep the stem of the funnel *pointed up*, and hold the stopcock as illustrated so that it neither opens nor falls out. Release the building pressure intermittently by opening the stopcock—between shaking, of course. You are, by shaking the system, producing a temporary emulsion and accelerating the attainment of an equilibrium between the ether and water phases of the system. After mixing in this way for a couple of minutes, replace the funnel in the ringstand, as shown in Figure 18.2, and allow the layers time to separate completely. The ether layer will be the top layer since the density of ether is only 0.7138 g/cm^3.

Remove the stopper from the funnel and carefully manipulate the stopcock to separate the bottom aqueous layer from the ether layer. Collect the aqueous layer in a clean receptacle. Transfer the ether layer from the funnel into another flask for the time being, but don't discard it or forget what it is. Cover this ether solution loosely with a watch glass or beaker. Return the aqueous layer of the amine hydrochloride to the empty funnel for washing.

Add 10 mL of cold fresh ether to the aqueous layer in the funnel, and shake as before to extract remaining portions of the benzoic acid and naphthalene from the aqueous layer. Allow the layers to separate and then again draw off the aqueous layer into a clean beaker. Label this, and leave the ether wash in the funnel.

To the aqueous solution of amine salt in the beaker, stir in 10% NaOH solution in small portions until the solution tests alkaline to litmus. Cool on

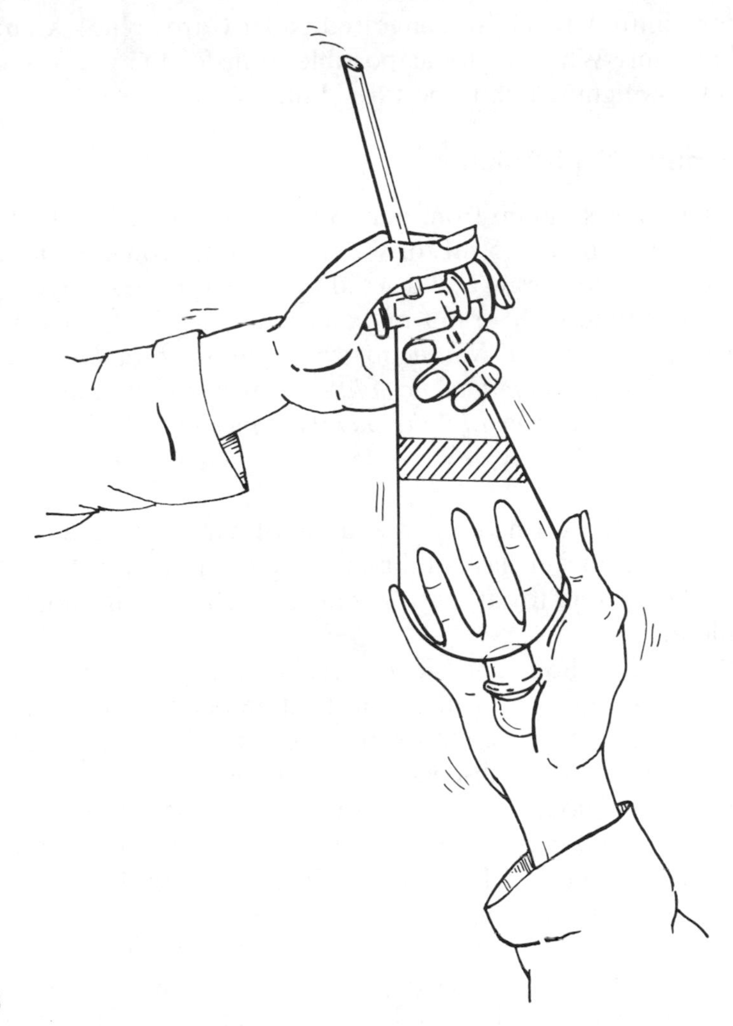

Figure 18.3

ice, and when precipitation is complete, filter off the regenerated *p*-nitroaniline on a small Büchner or Hirsch funnel with suction. Suck as dry as possible and transfer to a labeled weighing box for turning in.

B. Isolation of Benzoic Acid

Add the ether layer you removed in part A to the ether wash remaining in the separatory funnel. This solution should hold most of the benzoic acid and the naphthalene. To this ethereal solution add 25 mL of *cold* 5% NaOH. As before, shake and release pressure alternately until extraction has been maximized (about 2 minutes). Allow the layers to separate. The water layer now contains the water-soluble sodium benzoate, and the top ether layer contains naphthalene. Draw the aqueous sodium benzoate solution into a clean beaker and transfer the upper ether layer into another container for the time being. Cover it and don't lose track of it. Return the aqueous layer to the separatory funnel and, as before, wash this with 10 mL of fresh ether. Draw this washed aqueous layer of the sodium benzoate into a clean beaker and leave the ether extract in the funnel.

To the aqueous solution of sodium benzoate, stir in 10% HCl in small portions until it is acid to congo red paper (turns blue). Cool on ice and suction filter. When as dry as possible, transfer the recovered benzoic acid to a labeled weighing box to be turned in.

C. Isolation of Naphthalene

Add the ether solution from part B to the 10 mL ether wash in the separatory funnel and then add 15 mL of water. Shake thoroughly and allow the layers time to separate. Draw and discard the aqueous wash layer. Wash a second and a third time with 15 mL of water and each time discard the water layer.

This last extraction is done to remove most of the last traces of salts contaminating the naphthalene solution. All three of the washings in this procedure are mandated by the fact that ether and water do indeed have about a 5% mutual solubility and are never completely separated with a separatory funnel.

Transfer the washed ether solution of naphthalene to a dry, loosely covered Erlenmeyer flask and add 2–3 g of anhydrous $CaCl_2$. Slosh gently from time to time for at least 10 minutes. The drying should then be complete.

Ordinarily, the naphthalene is recovered by distilling off the solvent ether. If your lab is equipped with steam heat and enough hoods, your instructor may ask you to do this. However, ether fires are a potential danger with this method. Thus, in the absence of intructions to the contrary, simply pour about 5 mL of your dried ether solution into a clean, dry evaporating dish (or watch glass) and deposit the rest of the solution in the container labeled ETHER-NAPHTHALENE. (Your instructor can then recover the ether and naphthalene later and under safer conditions.)

Blow gently into the evaporating dish to evaporate the ether. Note the residue of naphthalene that remains. Note its odor and color. Transfer the dried naphthalene to a labeled weighing box. (Naphthalene is a white crystalline solid. Your product will probably have a yellow tinge due to the presence of lingering traces of *p*-nitroaniline. This demonstrates quite clearly the real difficulties attending *complete* separation of mixtures and purification of products. Chemically pure organic compounds are understandably expensive.)

Clean and return equipment to the storeroom and present your labeled *p*-nitroaniline, benzoic acid, and naphthalene to your instructor for approval.

DATA SHEET

Name: ______________________ Lab Section: ______ Date: ____________

THE SEPARATION OF A MIXTURE BY EXTRACTION TECHNIQUES

Properties of the compounds in the mixture (literature):

Compound	Structural Formula	Formula Weight	Melting Point	Solubility in Water
p-Nitroaniline	________	________	________	________
Benzoic acid	________	________	________	________
Naphthalene	________	________	________	________

Other observations:

QUESTIONS

1. In this experiment you recovered *p*-nitroaniline from the first (acidic) aqueous extract by making the solution alkaline with 10% sodium hydroxide. Write the equation for the neutralization reaction that precipitated the *p*-nitroaniline.

2. You recovered benzoic acid from the second (alkaline) aqueous extract by acidifying with hydrochloric acid. Write the equation for the neutralization that precipitated the benzoic acid.

3. In this experiment we conveniently separate a mixture of solids. Had our three-component mixture been composed of a liquid amine, a liquid acid, and a liquid neutral compound, how could these be isolated from their final solutions?

4. The solubility of benzoic acid in water at room temperature (25° C) is reported to be 3.4 g per liter. At 0° C it is 1.7 g per liter.

   a. How many grams of benzoic acid could be saved by cooling your last 25 mL of precipitated benzoic acid from room temperature to 0° C on ice before filtering?

b. On the assumption that you had 2.0 g of benzoic acid in the mixture you separated, what is the percent loss of yield one accepts by failing to cool the precipitated benzoic acid from 25° to 0° C on ice before filtering?

5. We used ether as the organic solvent. Many other organic liquids could be substituted for ether. Suppose that you were to select such an alternate solvent. List at least four qualities you should look for in the selection of the most promising alternate extraction solvent.

EXPERIMENT 19

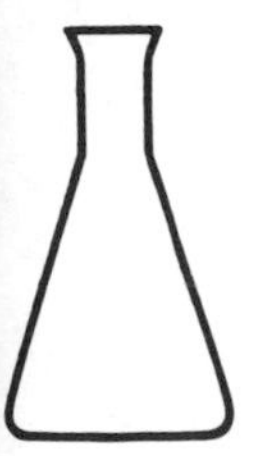

tert-Butyl Alcohol: Dehydration and Substitution Reactions

OBJECTIVES

1. To illustrate how a dehydration reaction might be accomplished.
2. To illustrate an organic substitution reaction and how it may be used to convert alcohols to halides.
3. To know how to calculate theoretical yields and percent yields.
4. To demonstrate at least four tests that distinguish saturation from unsaturation.

PRELABORATORY QUESTIONS

1. How does a dehydration reaction differ from a substitution reaction?
2. What is meant by "theoretical yield"?
3. Write equations to illustrate:
   a. protonation of isobutyl alcohol
   b. dehydration of isobutyl alcohol
   c. reaction of isobutyl alcohol with concentrated HBr

DISCUSSION

The reactions of alcohols are of paramount importance to chemists and physiologists. A somewhat atypical alcohol is selected for study in this exercise because its reactions are rapid. *tert*-Butyl alcohol, a tertiary alcohol, is easily subjected to dehydration and to substitution reactions. We shall do both and then, incidental to these reactions, have a look at some of the chemical properties of alkenes.

Consider the structure and predictable chemical properties of *tert*-butyl alcohol. The three electron-repelling alkyl groups endow the carbinol carbon with enough negative character that the OH group is easily lost to form a carbonium ion, especially after it has been protonated by an acid. In short, the intermediate *t*-butyl carbonium ion is a relatively stable entity and is therefore easily formed. As the following equations illustrate, the intermediate carbonium ion can then either pick up a chloride ion (substitution) or stabilize itself by giving up a proton (dehydration). By regulating reaction conditions, the chemist (you) can pick the options for the carbonium ion.

$$\underset{t\text{-Butyl alcohol}}{(CH_3)_3C{-}OH} \underset{\text{protonation}}{\overset{H^+}{\rightleftharpoons}} \underset{t\text{-Butyl oxonium ion}}{\left[(CH_3)_3C{-}OH_2^{\ +}\right]} \underset{H_2O}{\overset{-H_2O}{\rightleftharpoons}} \underset{t\text{-Butyl carbonium ion}}{\left[(CH_3)_3C^+\right]}$$

$$(CH_3)_3C^+ \underset{+H^+}{\overset{-H^+}{\rightleftharpoons}} \underset{\text{isobutene}}{CH_2{=}C(CH_3)_2}$$

$$(CH_3)_3C^+ \underset{-Cl^-}{\overset{+Cl^-}{\rightleftharpoons}} \underset{t\text{-Butyl chloride}}{(CH_3)_3C{-}Cl}$$

Keep well in mind that all of the steps indicated are reversible and that only by removing the ultimate product as it is formed can we hope to approach a quantitative reaction. This equation also graphically illustrates another common problem in organic synthesis, that of competing reactions. By carefully manipulating the reaction conditions—as we will try to do today—it is possible to guide the reaction course, for the most part, to the product we want. The 100% yield phenomenon, however, is almost exclusively the domain of *inorganic chemistry*, where ions are the stable entities and not merely transient intermediates.

Study the mechanism illustrated above in the equation. Review your classroom studies of dehydration and substitution reactions. While doing these syntheses, try to think of molecules at the angstrom level, observing their reactions from within the chemical system as it were. Think of the oxalic acid molecule as a proton donor and dehydration agent (it is indeed a good dehydrating agent when dry). Imagine the alternatives for an overpopulation of Cl^- ions. Protons don't really want them because HCl is a *strong* acid. What would you do in such a predicament? A good organic chemist can see molecules in his mind's eye and can have a "molecule's eye view" of their actions. When a chemist is able to do this, mechanisms become the laws under which molecules live and are indispensable aids to the chemist.

Review also the aqueous oxidation of organic compounds. This experiment can teach you a great deal about organic synthesis, mechanisms, and chemical testing. Your instructor may easily expand the exercise and give you two periods for its completion, or he or she may permit you to work in pairs so you have time to put more thought into it.

Wear safety glasses throughout this exercise!

MATERIALS NEEDED

Bring a *dry* graduated cylinder to the storeroom and get 20 mL of *tert*-butyl alcohol (enough for parts A and B of this experiment). Pick up also a 125 mL separatory funnel. The necessary concentrated HCl, anhydrous $CaCl_2$, and test reagents are available in the laboratory.

EXPERIMENTAL PROCEDURE

A. The Dehydration of *tert*-Butyl Alcohol

Set up the apparatus as illustrated in Figure 19.1. *Do not wash the distilling flask at this time!* In this, as in so many organic reactions, water is more troublesome than most of the impurities it is used to remove. The thermometer in the water bath is optional. It is not essential to the reaction, serving only to help you establish an optimum reaction temperature.

Prepare in advance labeled test tubes containing the test reagents 1 through 5 listed on the next page. If you do not do this in advance, your generator will become exhausted before you can complete the tests.

Weigh approximately 10 g of anhydrous oxalic acid on the top-loader or triple-beam balance and mix it with 10 mL of *tert*-butyl alcohol in the distilling flask. Secure all connections such that no gas leakage can occur.

Heat the water bath until a steady effervescence in the reaction flask indicates that dehydration is well under way. Allow the dehydration to procced for a couple of minutes to flush air from the system. Then run the following test reactions by bubbling the generated isobutene through the following (proceed rapidly from tube 1 through 2, 3, and 4 to 5, pausing only

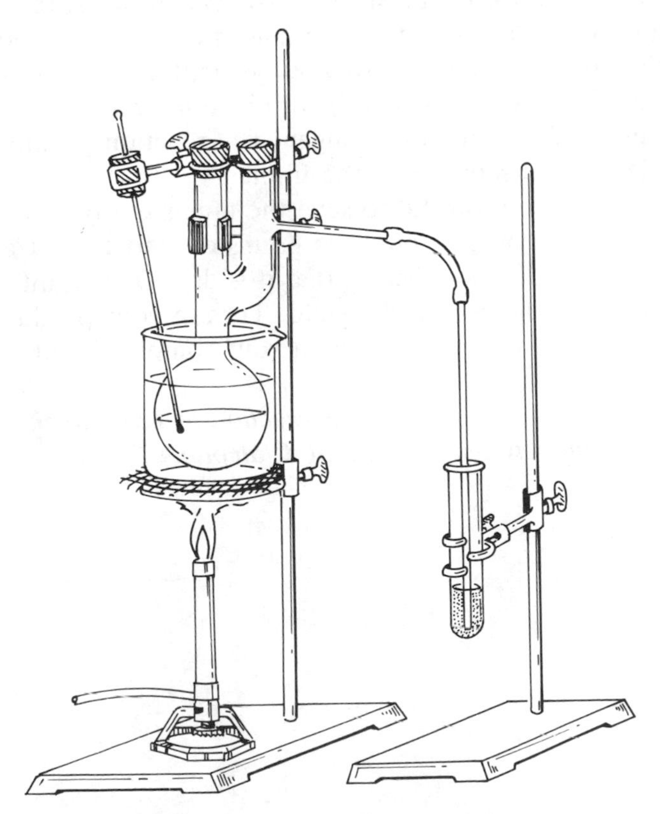

Figure 19.1

long enough with each to observe a change):

1. 1 mL of *concentrated H_2SO_4* in a small test tube: Record your observations and write the equation for the reaction on the data sheet.
2. 1 mL of *bromine water*: Observe, record, and write the equation for the reaction (see Experiment 15) on your data sheet.
3. 1 mL of *Baeyer's reagent*: Record your observations and write the equation. (See Experiment 15.)
4. A mixture of $\frac{1}{2}$ mL of 10% $K_2Cr_2O_7$ and $\frac{1}{2}$ mL of concentrated H_2SO_4 (*acidic dichromate*): Record your observations and write the equation.
5. 1 mL of *concentrated HCl*: Record your observations and write the appropriate equation.

When finished, flush all reaction mixtures down the sink.

B. The Synthesis of *tert*-Butyl Chloride

Obtain a 125 mL separatory funnel, check its stopcock lubrication, and set up in a ringstand, as illustrated in Figure 18.2. Carefully combine 10 mL of *tert*-butyl alcohol and 25 mL of concentrated HCl in the funnel. Secure both stopcock and stopper and mix slowly, with frequent release of pressure through the upturned stem. (See Figure 18.3 for the proper manipulation of a separatory funnel.)

If you fail to keep the stopcock securely in place and spill concentrated HCl on your hands or clothing, **wash immediately with a great excess of water!** Gradually increase the vigor of the shaking—between pressure releases—until 5 minutes have elapsed. After 5 minutes of reaction, add a couple of grams of anhydrous $CaCl_2$ as a dehydrating agent and as a source of extra Cl^- ions. Continue the reaction as before with careful and intermittent shaking and pressure release for 10 minutes longer. The reaction should then be complete, and two layers will separate in the funnel.

Separate the layers, being very careful to save the upper *tert*-butyl chloride layer. Dry this product over a couple of grams of anhydrous $CaCl_2$ until the disappearance of cloudiness indicates that it is dry. Decant into a preweighed vial, weigh again, and record the yield. Turn in your product, properly labeled, to your lab instructor along with your completed data sheet. Don't forget to complete the data table.

(This synthesis is a modification of the Lucas reaction of Experiment 20, a test reaction for distinguishing between 1°, 2°, and 3° alcohols.)

DATA SHEET

Name: ______________________ Lab Section: ______ Date: ____________

tert-BUTYL ALCOHOL: DEHYDRATION AND SUBSTITUTION REACTIONS

A. The Dehydration of *t*-Butyl Alcohol

Equation for the reaction:

Tests on isobutene:

1. Concentrated H_2SO_4 — Observations:

   Equation:

2. Br_2/H_2O — Observations:

   Equation:

   Baeyer's reagent — Observations:

   Equation:

4. $Cr_2O_7^{2-}/H^+$ — Observations:

   Equation:

5. Concentrated HCl — Observations:

   Equation:

B. The Synthesis of *t*-Butyl Chloride

Reactant or Product	Formula Weight	Grams Used or Yielded	Moles Used or Yielded	T_b or T_m	Density	Miscellaneous
$(CH_3)_3COH$	74			83° & 25°	0.779	∞ soluble in H_2O
HCl (conc)	36.5	25	0.307	110°	1.18	38% HCl by wt
$(CH_3)_3CCl$	92.6			51–52°	0.847	insoluble in H_2O

Theoretical yield: ____________________ Based on: ____________________

Actual yield: ____________________ Percent yield: ____________________

QUESTIONS

1. Write equations to show how you could convert

   a. cyclohexanol, C_6H_{11}—OH, to cyclohexene,

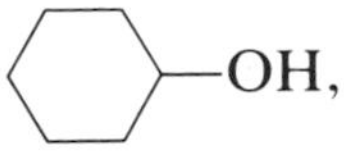

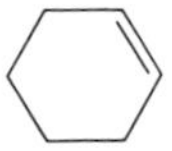

   b. cyclohexanol to cyclohexyl bromide (bromocyclohexane):

2. Consider this balanced equation for the oxidation of 2-butene to acetic acid with acidic dichromate:

$$3\ CH_3CH{=}CHCH_3 + 4\ Cr_2O_7^{2-} + 32H^+ \longrightarrow 6\ CH_3CO_2H + 8\ Cr^{3+} + 16\ H_2O$$

   a. How many grams of acetic acid, CH_3CO_2H, could theoretically be prepared from two moles of 2-butene?

   b. If the chemist running the reaction described in question 2a were able to isolate 150 g of acetic acid in a pure state, what was the percent yield?

   c. The chemist running the reaction above used $K_2Cr_2O_7$ as his source of dichromate ion. How many grams of $K_2Cr_2O_7$ would have been needed per liter of 2-butene gas? That is, how many grams of $K_2Cr_2O_7$ would have to be used to oxidize 1 liter (STP) of 2-butene to acetic acid?

3. Write an equation to illustrate the dehydration of isopropyl alcohol (2-propanol) with concentrated sulfuric acid.

4. Explain how you might distinguish between cyclopentane and cyclopentene with simple laboratory tests. (Describe what you would observe in any tests used.)

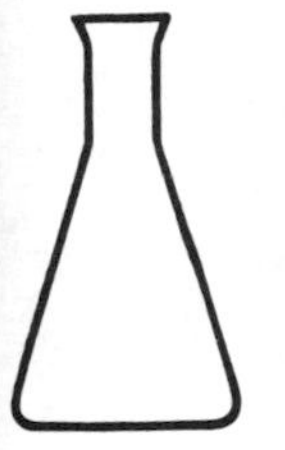

Organic Compounds Containing Oxygen: The Identification of an Unknown

OBJECTIVES

1. To be able to distinguish between the classes of organic oxygen-containing compounds from written structures.
2. To improve your command of organic nomenclature.
3. To be able to utilize lab techniques and tests for the identification of simple organic compounds containing oxygen.
4. To extend your proficiency with secondary reference sources.

PRELABORATORY EXERCISE

On the question sheet for this experiment, space has been provided for the inclusion of the names and structures of the 37 possible oxygen-containing compounds from which your unknown has been selected. Your instructor will provide these to you. Complete this part of your question sheet before starting your laboratory investigation. The limits within which these 37 compounds fall are specified in the following discussion. This preliminary preparation is designed to make the lab clear and meaningful.

DISCUSSION

With oxygen as an added building block, a great many new variations in the molecular architecture of organic molecules are possible, and new classes of compounds result. Seven of these classes are intimately involved in biological life and are characterized in this experiment. These are the alcohols, acids, esters, 1° amides, aldehydes, ketones, and ethers—all with unique structures found in biochemistry.

In this experiment, you will obtain an unknown compound and identify it. All unknowns issued for this investigation will be *monofunctional* compounds having *four or fewer carbon atoms. No unsaturated or cyclic* compounds are included. Furthermore, any amides will be *primary amides*. As a prelab exercise you have been asked to write the name and structure for each of these 37 compounds.

MATERIALS NEEDED

A vial of an unknown organic oxygen-containing compound. All other reagents, reference compounds, and materials needed are provided in the laboratory.

EXPERIMENTAL PROCEDURE

Examples from each compound class will be available in the laboratory for use in control tests. If you have done Experiment 16, the basic approach is the same: a process of elimination. Boiling points and densities, determined as described in Experiment 16, are your best means of establishing exact identification after you have narrowed the field to a class. The following classification tests are arranged in a practical order. It is prudent to follow this order in testing your unknown since many unnecessary tests and even hazardous reactions are thereby avoided.

A. Acids: RCOOH

These are water soluble when they contain no more than five carbons. There are five organic acids of four or fewer carbons considered in this experiment.

Litmus Test Moisten a strip of blue litmus paper with distilled water and place a drop (or a crystal) of the compound to be tested on the moist litmus. A red spot develops if the compound tested is a carboxylic acid. Since small amounts of acidic impurities are sufficient to give this test, it should always be confirmed with the bicarbonate test, which follows.

Reaction with Bicarbonate Carboxylic acids react with HCO_3^- to release carbon dioxide gas:

$$RCOOH + HCO_3^- \longrightarrow RCOO^- + [H_2CO_3] \longrightarrow CO_2\uparrow + H_2O$$

To 1 mL of 5% $NaHCO_3$ in a test tube, add one drop (or a crystal) of the compound to be tested. Watch for effervescence. Very weak acids such as phenols will give the litmus test but do *not* release CO_2 from a bicarbonate.

The Determination of Neutralization Equivalent Although this determination is optional for this experiment, it is the best and most precise measurement one can take of an acid. As was done in Experiment 8, a weighed sample of the acid is titrated to a phenolphthalein end point with standard base. By definition, then:

$$\text{neutralization equivalent} = \frac{g_{acid} \times 1000}{mL_{base} \times N_{base}}$$

If indeed you wish to try this determination, ask your instructor for some standard base, a buret, and a bottle of phenolphthalein.

B. 1° Amides: $RCONH_2$

These too are water soluble when of fewer than five carbons. In this experiment, we consider only the 1° amides of the five carboxylic acids from above.

Primary amides will release ammonia when heated with strong base:

$$RCONH_2 + NaOH \xrightarrow{\text{heat}} RCOONa + NH_3\uparrow$$

Add a couple of drops (or crystals) of the compound to be tested to 1 mL of 20% NaOH in a small test tube and heat gently. Remove from the source of heat and waft the vapors from the test tube toward your nose—carefully. The odor of NH_3 is easy to detect, but if you do not trust your nose, try holding a slip of *moist* red litmus over the mouth of the tube. Do not touch the tube with the paper since it is most probably strongly basic from the NaOH. Ammonia vapors will turn moist red litmus blue.

C. Aldehydes: RCHO

As are all oxygen-containing compounds having five or fewer carbons, these small aldehydes are water soluble. The aldehydes are easily oxidized (they are good reducing agents) and can be detected by a number of classification reagents. Two of these, the Tollen's test and the Schiff's test, follow.

Tollen's Silver Mirror Test Tollen's reagent must be prepared freshly and just before it is to be used. To prepare it, place a milliliter or two of NaOH solution (5%, 10%, 6 N) in a test tube and boil gently to thoroughly clean the inside walls of the tube. Pour the hot NaOH solution into the sink—*but do not rinse the test tube with water!* To the test tube than add approximately 2 mL of 2% $AgNO_3$ solution. A dirty brown precipitate of AgOH and its anhydride, Ag_2O, forms:

$$AgNO_3 + NaOH \longrightarrow AgOH\downarrow + NaNO_3$$
$$AgOH \longrightarrow \tfrac{1}{2}Ag_2O\downarrow + \tfrac{1}{2}H_2O$$

Dissolve this brown precipitate by adding dilute NH_3 dropwise and with shaking until the precipitate has dissolved and a clear water-white solution results. This is your Tollen's reagent. Be careful not to use more NH_3 than is absolutely necessary to dissolve the precipitate.

$$AgOH + 2NH_3 \longrightarrow \underset{\substack{\text{Ammoniacal silver hydroxide} \\ \text{(Tollen's reagent)}}}{Ag(NH_3)_2{}^+OH^-}$$

To the Tollen's reagent just prepared, add a couple of drops of the compound to be tested, mix, and watch for the formation of a mirror of free silver as it plates out on the inner walls of the test tube. If nothing happens, dip the tube into a beaker of gently boiling water—but *do not* heat with a direct flame. The oxidation-reduction reaction between an aldehyde and Tollen's reagent is:

$$RCHO + 2Ag(NH_3)_2^+ + 2OH^- \longrightarrow RCOO^-NH_4^+ + 2Ag\downarrow + 3NH_3 + H_2O$$

Six other compounds from the list of 37 studied in this experiment are also capable of reducing Tollen's reagent. These are: formic acid, $HCOOH$; the four formate esters, $HCOOR$; and formamide, $HCONH_2$. A close look at these structures will reveal that they possess the $\underset{\displaystyle H}{\overset{}{—\underset{|}{C}{=}O}}$ group just as the aldehydes.

Schiff's Test for Aldehydes Schiff's reagent is a fuchsia-colored dye that has been reduced to a colorless (leuco) state with bisulfite. When mixed with an aldehyde, the aldehyde absorbs the bisulfite, the dye is released, and its purple color reappears. This test is very sensitive, so be especially careful to avoid contamination of the reagent or your sample. If you get the reagent on your hands or clothing, persistent purple spots inevitably develop.

To test your unknown, simply add about a milliliter of the colorless Schiff's reagent to a *clean* test tube, and to this add a drop of the compound to be tested. *Do not heat!* The color should be fuchsia (purple). A pink color is negative. Try a control test.

D. Ketones: $R—\overset{\displaystyle O}{\overset{\|}{C}}—R$

These are water soluble when they contain fewer than five carbons.

The Iodoform Test for Methyl Ketones This test is not specific for all ketones, but since all noncyclic ketones of under five carbons *must* be methyl ketones, the test is significant for this experiment.

Other compounds among the 37 possible unknowns that give positive haloform reactions are: ethanal, ethanol, 2-propanol, and 2-butanol.

Place two or three drops of a methyl ketone (or unknown) in a clean test tube and add about 1 mL of 10% NaOH solution. To this add small portions of iodine in potassium iodide solution and shake intermittently. If the brown I_2 color fades away, it indicates that the organic compound is reducing it. If after several additions a pale yellow precipitate (finely divided at the start—milky and colloidal) comes down, the precipitate is iodoform, CHI_3, and indicates the presence of a methyl carbonyl compound or some compound easily oxidized to an intermediate methyl carbonyl compound. The active reagent in this test is sodium hypoiodite, NaOI, which forms in the equilibrium reaction between I_2 and NaOH:

$$I_2 + 2NaOH \longleftarrow NaOI + NaI$$

The iodoform test reaction can be generalized:

$$\underset{\text{a methyl ketone}}{\overset{\overset{\large O}{\|}}{RC}-CH_3} + 3NaOI \longrightarrow \overset{\overset{\large O}{\|}}{RC}-ONa + \underset{\substack{\text{iodoform}\\\text{(yellow ppt)}}}{CHI_3\downarrow} + 2NaOH$$

Some compounds are capable of reducing large quantities of NaOI, so be patient. Continue with the additions of I_2/KI until the brown color of I_2 persists. Also, an alkaline medium is necessary for the test, so if many additions of I_2/KI have been made, it is well to test the reaction mixture with red litmus paper to make sure that it is still alkaline. If not, add more 10% NaOH.

If, after the addition of large amounts of I_2/KI, a persistent brown color exists and the solution is alkaline, and there is *no* yellow precipitate, the test is *negative*. You know, then, that your compound indeed has a reducing capacity but it is *not* a methyl ketone, ethanol, acetaldehyde, isopropyl, or *sec*-butyl alcohol since no iodoform formed.

E. Alcohols: 1°, RCH_2OH; 2°, R_2CHOH; and 3°, R_3COH

Alcohols containing fewer than five carbons are also water soluble by virtue of their H-bonding capabilities. Three tests are listed. All three tests distinguish qualitatively between 1°, 2°, and 3° alcohols, but they are not precise and are easily misinterpreted. For this reason, it is prudent to run all three before drawing a final conclusion. Remember that alcohols are *neutral* compounds. One should never run these tests on acidic or basic compounds or on compounds known to be amides, aldehydes, or ketones. If you ran tests in the order presented in this experiment, you will know whether tests for alcohols are warranted.

The Lucas Test for Water-Soluble Alcohols **CAUTION:** Lucas reagent is a very strong acid. It is concentrated HCl saturated with the strong Lewis acid, $ZnCl_2$. Lucas reagent will convert alcohols to their corresponding chlorides as follows:

$$\underset{\text{soluble alcohol}}{ROH} + HCl/ZnCl_2 \longrightarrow \underset{\text{insoluble halide}}{RCl\downarrow} + H_2O + ZnCl_2$$

A cloudiness develops as the insoluble halide forms; this is the basis for interpreting the test. Time the appearance of the cloudiness. In general,

3° alcohols react in less than 5 minutes.

2° alcohols react in 5–15 minutes.

1° alcohols react in more than 15 minutes.

To run the Lucas test, add four drops of the liquid alcohol to be tested to 2 mL of Lucas reagent in a *dry* test tube. Shake gently to mix, stopper, and allow to stand at room temperature. Time the appearance of a cloudiness, if any develops.

The Metallic Sodium Test for Alcohols CAUTION: Sodium reacts violently with water. Keep everything scrupulously dry and wear your safety goggles! In contrast with the Lucas test reagent, metallic sodium reacts faster with 1° alcohols and slowest with 3° alcohols. Remember also that the presence of *any* water is not only dangerous but it invalidates the test. *Use dry equipment!* Watch for the evolution of H_2 gas:

$$2ROH + 2Na \longrightarrow 2RONa + H_2 \uparrow$$

1° alcohols give immediate and rapid bubbling.

2° alcohols give slow, controlled bubbling.

3° alcohols give very slow, almost imperceptible bubbling.

If the evolution of hydrogen gas is noticed in the beginning but then slows and stops, the presence of traces of water is strongly suggested. With 3° alcohols, it is usually observed that tiny bubbles slowly form on the surface of the sodium and that tapping the tube gently will release them. **Discard waste sodium-alcohol in the waste sodium jar—and replace the cover loosely.** ***Do not dump in the sink!***

The Chromic Oxide Test for 1° and 2° Alcohols CAUTION: This reagent is a strong acid and a powerful oxidizing agent! The test depends on the resistance of 3° alcohols, esters, and ethers to oxidation. Primary and secondary alcohols are quickly oxidized, and an opaque blue-green suspension of a Cr^{3+} complex forms as the oxidation-reduction proceeds:

$$3R_2CHOH + \underset{\text{orange-brown solution}}{2CrO_3 + 6H^+} \longrightarrow 3R_2C{=}O + \underset{\text{opaque blue-green}}{2Cr_{3+}} + 6H_2O$$

To run the test, place 1 mL of acetone in a dry test tube and to this add one drop of the alcohol to be tested. Mix, add one drop of the CrO_3/H_2SO_4 reagent, and mix again. The rapid formation of an opaque blue-green color suggests the presence of a 1° or 2° alcohol if it forms *within 3 seconds*. Many compounds, including 3° alcohols, oxidize more slowly but ultimately produce the blue-green color. Aldehydes react rapidly with the reagent, but your earlier tests would have exempted its consideration as a possible alcohol.

F. Esters: RCOOR

Esters have fruity, pleasant odors. Like other oxygen-containing compounds, they are water soluble when they contain fewer than five carbons. In this experiment, we consider no fewer than seven esters: one propionate, two acetates, and four formates. Recall from part C (aldehydes) that formates will indeed give a positive Tollen's test. There are no good tests that are specific for esters, so you will have to depend on physical properties such as densities and boiling points, and upon the process of elimination to identify these. Be advised that esters hydrolyze slowly in the presence of moisture and therefore, when not freshly prepared and pure, will contain traces of the parent acid—enough to give a positive litmus test. Saponification equivalents best characterize esters, but we will not attempt them in this experiment. Saponification is the subject of Experiment 23.

G. Ethers: ROR

These are the most inert of the oxygen-containing compounds, and no *easy* definitive tests are available. They are volatile and highly flammable. There are five simple ethers of four or fewer carbons. Their identification is best accomplished through their physical properties and the process of elimination.

Table 20.1 may prove helpful in interpreting your test results. (Please do *not* assume, as the table suggests, that *all* tests are to be run on *all* classes of compounds.)

Determination of the Physical Properties of the Unknown

Having classified your unknown, it only remains to specifically identify it. Since the scope of this experiment is small and the number of compounds representing each class is very limited, the determination of a physical property or two is usually sufficient for conclusive identification.

Look up the boiling points, densities, and refractive indices for the compounds representing the class to which your unknown belongs. (Refer to your prelaboratory report.) Usually, the boiling points vary enough that the unknown is readily identified. In some cases, densities or refractive indices are definitive. Decide which physical property (or properties) will be most helpful to you and establish it (or them) for your unknown.

Procedures for the determination of boiling points and densities are described in Experiment 16. Your laboratory instructor will help you measure the refractive index if you conclude that a refractive index value is needed for an unequivocal identification.

Identify your unknown from the list of 37 possibilities in your prelaboratory exercise, and complete the data sheet.

TABLE 20.1 Expected Test Results for Low-Molecular-Weight Oxygen-Containing Compounds

	$RCOOH$	$RCONH_2$	$RCHO$	$RC(=O)CH_3$	RCH_2OH	R_2CHOH	R_3COH	$RCOOR$	ROR
Litmus	+	–	–	–	–	–	–	–(?)	–
HCO_3^-	+	–	–	–	–	–	–	–	–
NaOH/heat	–	+(NH_3)	–	–	–	–	–	–	–
Tollen's	–	–	+	–	–	–	–	–	–
Schiff's	–	–	+	–	–	–	–	–	–
Iodoform	–	–	+/–	+	+/–	+/–	–	–	–
Lucas	–	–	–	–	v slow	slow	fast	–	–
Metallic Na	+!	+/–	+/–	+/–	v fast	fast	slow	–	–
CrO_3/H_2SO_4	–	–	+	–	+	+	–	–	–

(These tests should not be tried)

DATA SHEET

Name: ______________________ Lab Section: ____ Date: __________

ORGANIC COMPOUNDS CONTAINING OXYGEN: THE IDENTIFICATION OF AN UNKNOWN

Unknown number: __________ Unknown is: ______________

Structural formula of unknown:

Physical properties of identified unknown (literature):

Tests run—with equations for positive tests:

Other observations and evidence for identity:

QUESTIONS

The list of unknowns from which yours was issued total 37 compounds. All are monofunctional, saturated, and noncyclic. All have four or fewer carbons. Write the structures for all 37 possibilities and name each by an acceptable system of nomenclature.

1. The five carboxylic acids are:

2. The five 1° amides are:

3. The five aldehydes are:

4. The two ketones are:

5. The eight alcohols are:

6. The seven esters are:

7. The five ethers are:

Separation of Food Colorings: Analysis for Vitamin C in Fluid

OBJECTIVES

1. To establish the relative retention capabilities of a disposable extraction column for various food colorings.
2. To use a disposable extraction column for decolorizing a solution in preparation for analysis by titration.
3. To use iodimetry in an analysis for vitamin C.

PRELABORATORY QUESTIONS

1. Define the terms *lipophilic* and *lipophobic*.
2. Reproduce the structure for Green 3 (Figure 21.1), drawing circles around all ionic sites and underlining all additional hydrophilic sites capable of H-bonding.
3. If the titration of vitamin C with iodine is done without the addition of ammonium acetate, the reaction will proceed normally at first but then rapidly slow down. Consider the nature of the reactants and the products and give a reason why the ammonium acetate facilitates the reaction.
4. What is the "common ion effect"?

DISCUSSION

Many food and hygiene products contain food colorings. These usually have no function other than to make the product more attractive and more marketable. Most food colorings are large organic molecules with sufficient ionic character to ensure water solubility. Some examples are seen in Figure 21.1.

There are two parts to this experiment. In part A you will remove food colorings from water solutions to familiarize yourself with the use of a

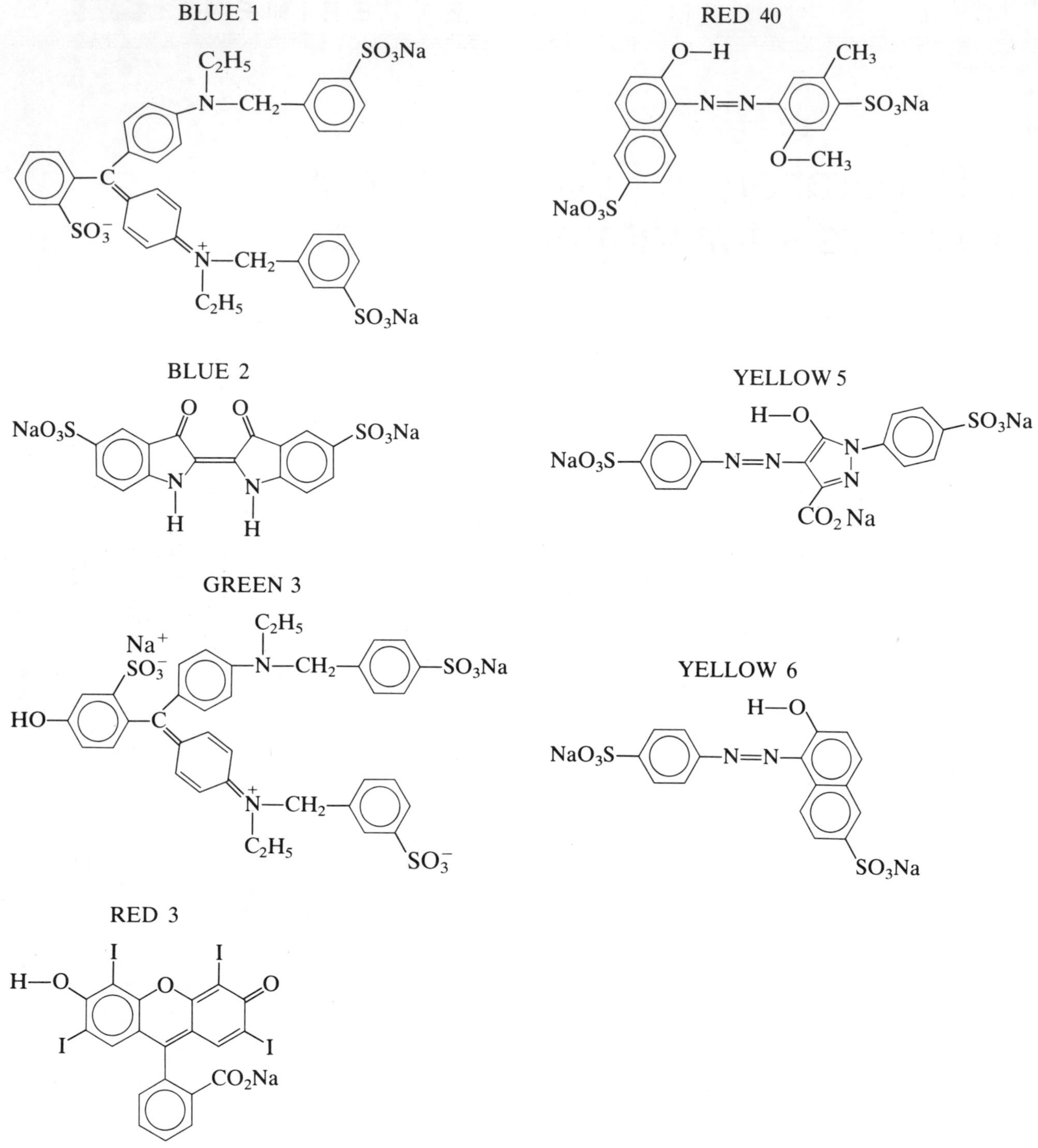

BLUE 1

RED 40

BLUE 2

YELLOW 5

GREEN 3

YELLOW 6

RED 3

Figure 21.1 Structures of FD&C Dyes

disposable extraction column. Then, in part B, you will use this extraction technique in the analysis of a vitamin C sample.

Disposable extraction columns like the C-18 SEP-PAK discussed below present a new dimension to extraction techniques (Figure 21.2). Until recently the only way to remove a *nonpolar* solute from a biological fluid was to

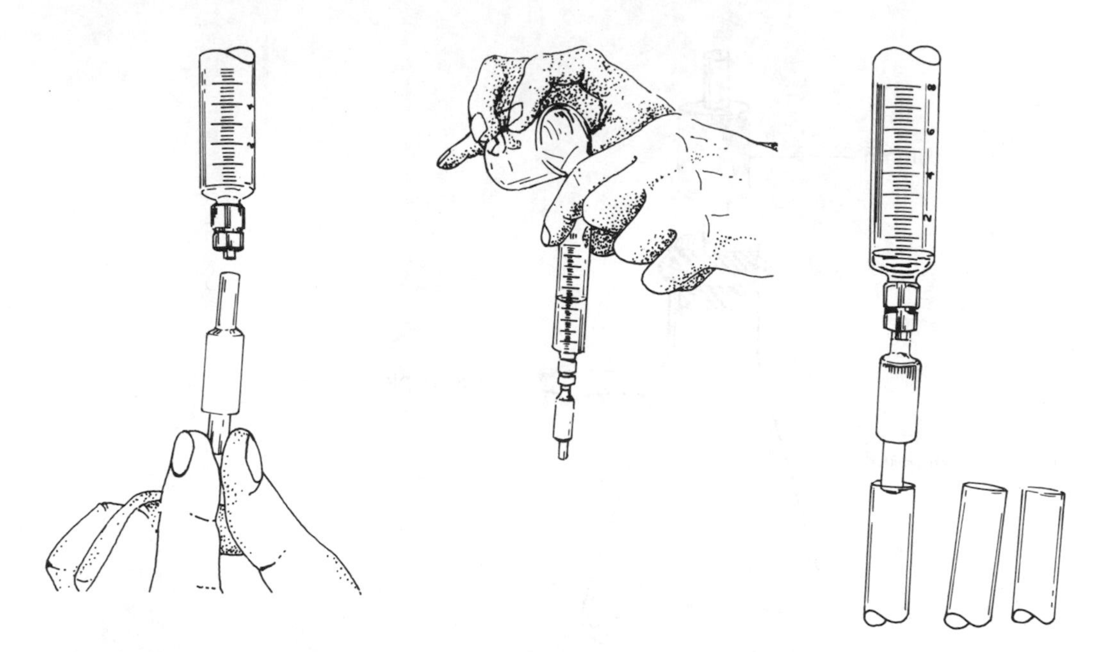

Figure 21.2

extract with an immiscible nonpolar solvent, as was done in Experiment 18. Results tediously obtained are often not quantitative because the solvent systems are so difficult to separate completely. The SEP-PAK is a small plastic column containing a nonpolar organic solvent covalently bound to a finely divided solid support. As a biological fluid passes through such a column, its nonpolar components are selectively absorbed while the polar (water-soluble) components flush on through. The absorbed nonpolar components are then easily washed free of remaining traces of polar material by simply purging the column with water.

To recover the absorbed material from the column, one first removes any adhering water by drying the column in a stream of air (see Figure 21.3). Once well dried, the absorbed material is flushed from the column (eluted) with an appropriate organic solvent. We will use isopropyl alcohol–water solvent pairs.

This method of extraction offers many advantages. It is rapid, requiring less than a minute per extraction. Annoying emulsions, so common with liquid–liquid extractions, are avoided. But perhaps the best feature is that simple water-soluble alcohols can be used as eluting solvents instead of the otherwise preferred—but biologically hazardous—chlorinated hydrocarbons. One last advantage is that with proper care the life of a disposable extraction column can be extended through repeated usage. We will demonstrate this. The column issued to you will be washed and dried and saved for possible use in Experiments 36 and 37.

A few simple sequencing rules must be observed, however, to ensure the best results:

1. Disposable extraction columns must be *primed* before use. This is done by flushing slowly with the elution solvent of choice, and then rinsing with distilled water. This process ensures that the column will extract with maximum efficiency.

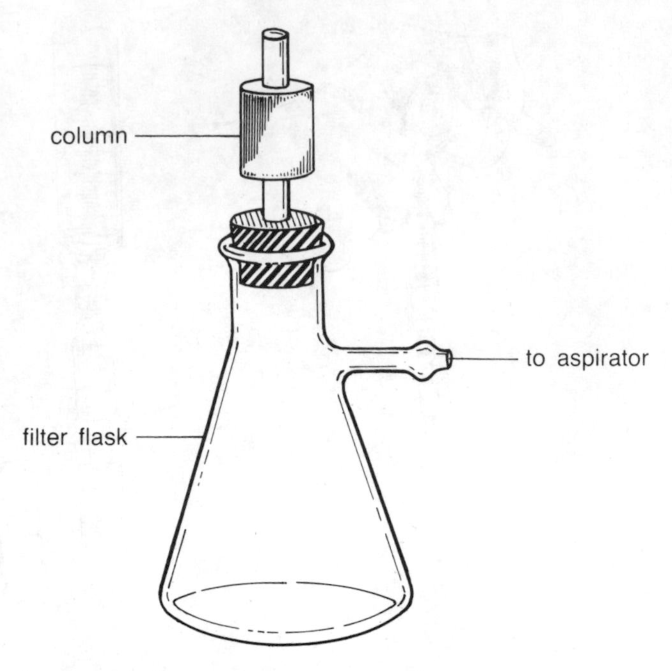

Figure 21.3

2. When extracting, rinsing, or eluting, pass the fluid through the column slowly, one or two drops per second.
3. Absorbed nonpolar materials should always be *washed* free of ionic impurities with water or with a specified alcohol–water solution before they are eluted. Ionic impurities can effectively interfere with both the elution process and the subsequent analysis.
4. *Dry* the column before eluting. This is easily done by aspirating (2 minutes or more) through a stoppered filter flask (Figure 21.3). Even small amounts of adhering water markedly reduce the elution potency of the solvent.

In summary, then, always prime, absorb slowly, wash, dry, and elute slowly—in that order.

It is also highly advisable to analyze your sample as soon as possible after elution. Biological materials, especially, have a propensity for decomposition.

The reaction between vitamin C (ascorbic acid) and 1% I_2 in methanol is quantitative and very rapid. If you did Experiment 17 you ran a very similar analysis for iodine. The only difference is that here ascorbic acid is the reducing agent instead of thiosulfate ion. Since the redox reaction is complete within 2 seconds, iodimetric titration is easily accomplished. The reaction occurs as follows:

$HOCH_2$–HOCH–(ring: O, C=O, HO–C=C–OH) $\underset{\text{ization}}{\overset{\text{tautomer-}}{\rightleftharpoons}}$ $HOCH_2$–HOCH–(ring: O, C=O, HO–C–H, C=O) $\underset{I_2}{\overset{\text{redox}}{\rightleftharpoons}}$ $HOCH_2$–HOCH–(ring: O, C=O, C=O, C=O) + 2HI

Ascorbic acid — Dehydroascorbic acid

Using starch as the indicator, one need only titrate to the appearance of the deep blue starch–iodine complex.

For your calculations, you will need to remember that the iodine solution is standardized such that 1 mL is equivalent to 6.9 mg of vitamin C.

MATERIALS NEEDED

A 30 mL unknown similated urine specimen containing ascorbic acid, a C-18 disposable extraction column (SEP-PAK), and a syringe for use with the SEP-PAK will be needed. Also needed will be a buret and a buret holder. All other materials will be available in the laboratory.

EXPERIMENTAL PROCEDURE

A. Determination of Relative Retentions of Food Colorings by a C-18 SEP-PAK

Obtain about 20 mL of each of the four solutions of basic food colorings provided.

1. Prime the SEP-PAK as described in the discussion section by rinsing first with 5 mL of methanol and following this with a water rinse.
2. Add 5 mL of one of the colored solutions to the syringe and force through the column at a rate of two drops per second into a test tube. Note (and record on the data sheet) whether the color is retained partially or completely by the column. At this point the absorbed material is usually washed free of water-soluble impurities. You may skip the "wash" here since there are no soluble impurities in these food coloring solutions.
3. **a.** *If retention is complete*, elute the column with 10% isopropyl alcohol–water (10 mL). From an examination of the column and the eluate determine whether or not elution was complete. If it is *not*, elute with 10 mL of 50% isopropyl alcohol–water. Record your observations.

   b. If the column *retention was not complete* in your original elution of the colored solution, you must start over. This time the lipophilic (fat-loving) character of the coloring molecules will be enhanced by adding brine. The common ion effect should inhibit ionization of the dye molecules (note structures in Figure 21.1) and render them less water soluble and more soluble in the hydrocarbon coating on the column packing. Four steps are then taken:

   (1) Clean the column by rinsing with 50% isopropyl alcohol–water. Ten mL eluted at a rate of two drops per second should suffice.

   (2) Reprime the column with 5 mL of methanol followed by a 5 mL water wash.

   (3) Add 2 mL of saturated NaCl solution (brine) to 5 mL of the colored solution to be tested. Mix and pass the resulting 7 mL through the column and into a test tube. Record on the data sheet your observations on the retention of the color by the column.

(4) Elute with 10 mL of water and record your observations. If color is retained on the column, elute again with 10 mL of the 10% isopropyl alcohol–water and record your observations. If color is still evident in the column, elute finally with 10 mL of the 50% isopropyl alcohol–water. Record all observations on the data sheet.

Repeat part A for the other colored solutions provided. From your observations, list on the data sheet the four colors in order of their *decreasing* retention by the column. You now have established the practical limits of the column and isopropyl alcohol–water systems for use with food colorings.

B. Vitamin C Determination

Obtain a 20 mL unknown specimen containing ascorbic acid and record its number. This unknown will be a similated urine specimen colored with a mixture of the food colorings studied in part A. In preparation for the iodimetric titration for vitamin C content, the specimen must first be decolorized. For this we employ the disposable column.

1. Reprime your column with 5 mL of methanol followed by a 5 mL water wash.
2. Use a pipet to measure exactly a 5.0 mL aliquot of the specimen into the syringe, and take special care to avoid spilling. Decolorize this specimen aliquot by forcing it through the column and *directly* into a clean 125 mL Erlenmeyer flask. Next, wash the column directly into the same Erlenmeyer flask with 10 mL of water. You have in this operation removed similated bile pigments from the ascorbic acid solution and have decolorized it for the titration with iodine.
3. To the decolorized eluant in the flask add ten drops of 1% starch indicator solution, 1 mL of 0.5 M ammonium acetate solution (a buffer), and about 25 mL of water. This water is added to dilute the methanol solvent of the standard iodine solution. (Too much methanol destroys the blue starch–iodine complex and makes the end point fleeting and vague.) Rinse down the flask walls with H_2O.
4. Fill a buret with the standard iodine solution provided. Be sure to rinse the buret with a few milliliters of the I_2 solution first. Record the strength of the I_2 solution.
5. Titrate your unknown ascorbic acid solution to a faint blue end point. As the end point is approached, the fading of the initial blue color slows down. A proper end point is one where one drop of the I_2 solution produces a blue tint that persists for 30 seconds or more. Record the volume of standard I_2 used.

Repeat part B with additional 5 mL aliquots of the specimen until you are satisfied that an average reading is accurate enough for the assay of the specimen.

DATA SHEET

Name: ______________________ Lab Section: ______ Date: __________

SEPARATION OF FOOD COLORINGS: ANALYSIS FOR VITAMIN C IN FLUID

A. Data

	← *Retention*	→	←	*Elution*	→
	Colored Solution Only	Solution with Brine	Water	10% Isopropanol	50% Isopropanol
Red					
Blue					
Green					
Yellow					

B. Data and Results

Strength of the standard I_2 solution: ______________________

Volume of vitamin C solution used: ______________________
(volume of aliquot titrated)

Titration Data

	Run 1	Run 2	Run 3	Run 4	Run 5
mL of I_2 solution used					

Average volume of I_2/methanol used: ______________________
Concentration of vitamin C (mg/mL) in specimen # ________ is __________

QUESTIONS

1. List the food colorings in order of decreasing retention on the column used.

2. Why does the addition of brine cause some colorings to adhere more strongly to the column?

3. Cold dilute methanolic I_2 is just strong enough as an oxidizing agent to oxidize ascorbic acid mole for mole to dehydroascorbic acid. (See the equation in the discussion section.) Moreover, only a slightly stronger oxidizing agent will convert ascorbic acid all the way to CO_2 and H_2O. Explain why ascorbic acid is effectively used by housewives when canning peaches or apricots to prevent darkening of the fruit.

4. What is the concentration of your unknown vitamin C solution in terms of ppm?

5. What is the molarity of your unknown vitamin C solution?

EXPERIMENT **22**

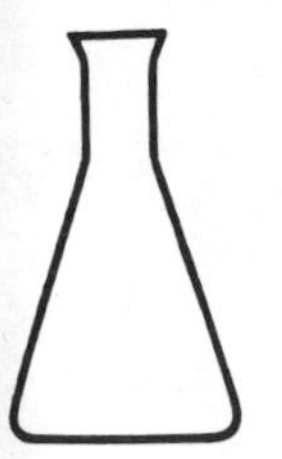

Esterification: Aspirin and Oil of Wintergreen

OBJECTIVES

1. To illustrate two simple organic esterification reactions.
2. To explore the physical and chemical properties of aspirin and oil of wintergreen.
3. To show how to distinguish between a carboxylic acid and a phenol.

PRELABORATORY QUESTIONS

1. Define esterification and write a general equation to illustrate esterification.
2. What is a phenol, and how does it differ from an alcohol?
3. "Soluble aspirin" is simply sodium acetylsalicylate. Draw a structure for "soluble aspirin" and explain why it is indeed more water soluble than ordinary aspirin.
4. Write an equation illustrating the reaction of salicylic acid with the strong base, NaOH.
5. Why is it advantageous to eliminate water during esterification?

DISCUSSION

Aspirin is the greatest of the wonder drugs. Discovered in 1899, it has a triple-threat medicinal action. Simultaneously, it serves as an analgesic (pain killer), an antipyretic (fever depressant), and an antiseptic (germicide). The daily consumption of aspirin in the United States alone exceeds 21 tons.

Aspirin (acetylsalicylic acid) is the simple ester of salicylic acid and acetic acid. The ordinary esterification process that forms aspirin proceeds rather slowly due to the acidic nature of the phenolic hydroxyl group. We shall speed the process considerably by using the more reactive acetic anhydride in place of acetic acid. The equation for the reaction is:

$$C_6H_4(OH)(CO_2H) + (CH_3CO)_2O \xrightarrow[\text{catalyst}]{H_2SO_4} C_6H_4(OCOCH_3)(CO_2H) + CH_3COOH$$

Oil of wintergreen (methyl salicylate) is another ester of salicylic acid. Methyl salicylate enjoys widespread use as a condiment and as a linament. This experiment includes an esterification of salicylic acid and methanol to form methyl salicylate:

$$C_6H_4(OH)(COOH) + CH_3OH \overset{H^+}{\rightleftharpoons} C_6H_4(OH)(COOCH_3) + H_2O$$

This "ordinary esterification"—the reaction between a carboxylic acid and an alcohol—amounts to a slowly attained equilibrium.

In this experiment you are to prepare and purify aspirin, run a few tests on the product, and then synthesize a sample of oil of wintergreen (methyl salicylate). Your instructor may ask you to work in pairs, with one of you synthesizing aspirin while the other synthesizes oil of wintergreen.

Ordinary (or simple) esterification can be described as the reaction between an acid and an alcohol that produces an ester and water. It is a reaction that proceeds *slowly* to a position of equilibrium in which the reverse hydrolysis reaction attains the same rate as the forward esterification reaction. This position of equilibrium commonly rests at about 75% product and 25% reactant:

$$RCO_2H + R'OH \underset{\text{hydrolysis}}{\overset{\text{esterification}}{\rightleftharpoons}} RCO_2R' + H_2O;\ K_{eq} \simeq 4$$

Catalytic esterification improves the speed with which equilibrium is reached but does not alter that equilibrium position. The maximum yields are still about 75%. The acid catalyst provides protons for the transformation of the reactant carboxylic acid into a better target for the alcohol (itself a Lewis base):

$$\mathrm{R\overset{O}{\overset{\|}{C}}{-}OH + H^+ \rightleftharpoons \left[R{-}C^+\begin{smallmatrix}OH\\OH\end{smallmatrix}\right] \underset{H}{\overset{R'{-}O}{\rightleftharpoons}} \left[R{-}\underset{OH}{\overset{OH}{C}}{-}O^+\begin{smallmatrix}R'\\H\end{smallmatrix}\right] \overset{-H^+}{\rightleftharpoons}}$$

Planar carbonium ion; Oxonium ion

$$\mathrm{\left[R{-}\underset{OH}{\overset{OH}{C}}{-}OR'\right] \xrightarrow{-H_2O} R{-}\overset{O}{\overset{\|}{C}}{-}OR'}$$

We will use acid catalysis in our synthesis of both aspirin and oil of wintergreen.

A study of the above equilibria shows that if indeed we could remove water from the reaction scene as it is formed, no reverse hydrolysis would be possible. Yields approaching 100% could then be expected. A great many ingenious techniques have been developed that do precisely that. We shall, however, try yet another scheme for maximizing yield in our synthesis of aspirin. We will substitute the anhydride of the carboxylic acid for the acid itself. In so doing, we eliminate the possibility of producing water as a product, and of course, the possibility of a competing reverse reaction as well. We shall also add a few protons to catalyze the reaction.

Esterification using an acid chloride (actually a mixed acid anhydride) will serve to illustrate how the proton assists (catalyzes) the reaction:

$$\mathrm{R{-}\overset{O}{\overset{\|}{C}}{-}Cl + H^+ \rightleftharpoons \left[R{-}C^+\begin{smallmatrix}OH\\Cl\end{smallmatrix}\right] \underset{H}{\overset{R'{-}O}{\rightleftharpoons}} \left[R{-}\underset{Cl}{\overset{OH}{C}}{-}O^+\begin{smallmatrix}R'\\H\end{smallmatrix}\right] \overset{-H^+}{\rightleftharpoons}}$$

$$\mathrm{\left[R{-}\underset{Cl}{\overset{OH}{C}}{-}OR'\right] \xrightarrow{-HCl} R{-}\overset{O}{\overset{\|}{C}}{-}OR'}$$

Note that in the last step there is no equilibrium; the reaction cannot be reversed once it gets to the final step.

MATERIALS NEEDED

You will need a reflux condenser with appropriate rubber tubing, a small filter flask and a Büchner or Hirsch funnel with paper to fit, a round-bottom flask (100 mL) if Erlenmeyer flasks are not used, and a small 50 or 100 mL separatory funnel. If available, heating mantles should be used to minimize the danger of fires. All other items of equipment and chemicals will be available in the laboratory.

EXPERIMENTAL PROCEDURE

A. Acetylsalicylic Acid

1. Weigh out 2 g of salicylic acid in a weighing box. (Figure 22.1 illustrates how to make a weighing box.) Transfer to a large, *dry* test tube and add 4 mL each of glacial acetic acid and acetic anhydride. The latter is an effective lachrymator so a hood should be used if available. **Wear safety glasses**! Stir this mixture with a glass rod until the solid has dissolved, and then stir in four drops of concentrated H_2SO_4. Place the test tube in a water bath (a 250 mL beaker half full of *gently* boiling water) and allow the esterification to proceed at 100° for 30 minutes.

   While this reaction is proceeding you might save time by starting with part B.

2. After a half hour, the aspirin synthesis should be complete. If you wish, you may heat it longer for possibly a better yield. To recover the aspirin from the reaction mixture, pour the hot mixture into a 150 mL beaker

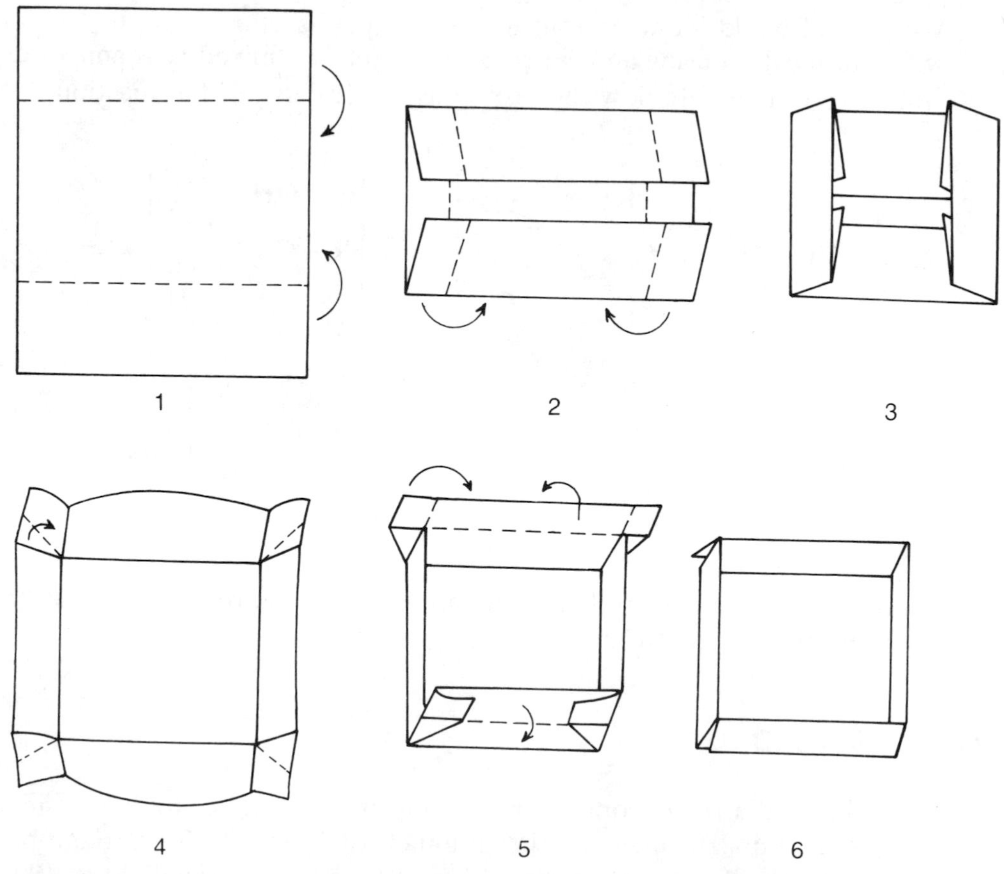

Figure 22.1

containing 10 g of crushed ice and 20 mL of water. Rinse the reaction tube with an additional 10 mL of cold water and add to the beaker. Stir the mixture until the ice has all melted and crush any lumps that form. If no solid has precipitated, you should cool the mixture by placing it in a larger beaker containing a crushed-ice slurry. Scratch the walls with a stirring rod to induce vibrations that can start crystal formation.

3. Collect the product by filtration on a Büchner or a Hirsch funnel under aspirator pressure (Figure 18.1). Rinse a couple of times with small amounts of cold water. First disconnect the aspirator hose, then add the water while gently triturating (mashing to a slurry) such that the filter paper is not damaged, and finally reconnect the aspirator and suck to dryness. When it is *as dry as possible*, transfer the cake to a weighing box. This can be done by first loosening the circumference of the cake with a spatula, as one might loosen a freshly baked cake from a tin, and inverting the funnel in the weighing box. Tap the inverted funnel, and the filter cake will deposit. The filter paper can then be peeled off. To accelerate drying, the product should be broken up and spread around the box. Proceed with the following tests while it dries.

4. a. *Test for phenols:* Salicylic acid, the starting material, is both an acid and a phenol. A very sensitive test for phenols involves adding a drop of ferric chloride solution. Minute traces of a phenol will produce a distinct color ranging from reds and violets to blues and greens, depending on the identity of the phenol. Run a control first by adding a single, tiny crystal of salicylic acid to a couple of milliliters of water in a test tube and then mixing in a drop of $FeCl_3$. Using a clean test tube, test your aspirin in the same way and interpret your results. If your aspirin is impure, you might want to recrystallize it. Question 2 on the question sheet describes a suitable procedure for recrystallizing aspirin.

b. *Test for acidity:* Stir a small portion of your aspirin into a milliliter or two of 5% sodium bicarbonate solution. Observe carefully. Place a crystal or two of your aspirin on a piece of *moist* blue litmus and observe its acidity. *Simple phenols are so weakly acidic that they will not react with bicarbonate. They are, however, acidic enough to give a positive litmus test.* Thus one can distinguish between a phenol and a carboxylic acid with these two simple tests.

When the aspirin is dry, weigh it and determine its yield, the theoretical yield, and the percent yield.

B. Methyl Salicylate

1. Assemble a reflux apparatus as illustrated in Figure 22.2 (a 100 mL round-bottom flask may be substituted for the 125 mL Erlenmeyer flask illustrated).

Weigh out 2.0 g of salicylic acid and mix this with 30 mL of methyl alcohol in the reaction flask. **Be careful to extinguish all fires in the vicinity first. Methanol boils at 65° and is highly flammable!** As soon as all of the salicylic acid has dissolved, add five drops of concentrated H_2SO_4 as a catalyst.

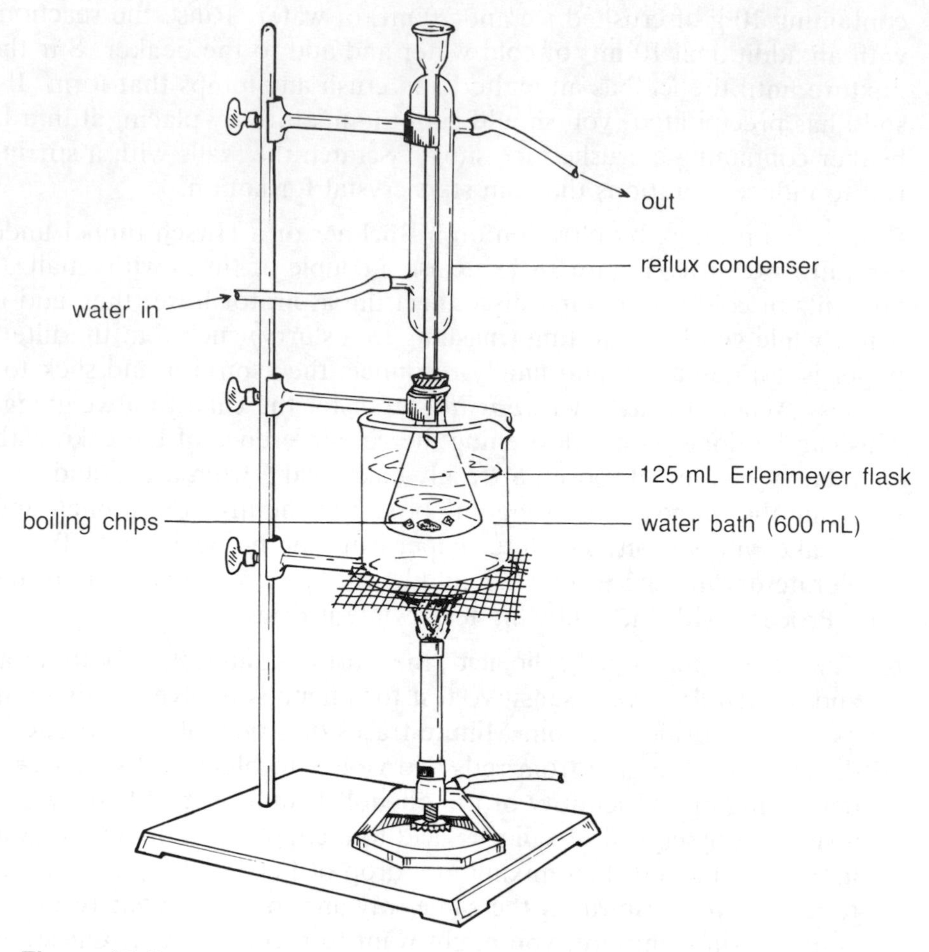

Figure 22.2

2. Secure the reflux condenser, turn on the cooling water to circulate at a moderate rate, add a couple of boiling chips, and heat the water bath until its temperature is between 65° and 70° C—just hot enough to very gently reflux the methanol in the reaction flask.

   Continue this gentle reflux for at least an hour. The longer the reaction time, the better your yield of methyl salicylate.

3. After an hour (or more) of refluxing, turn off the flame and the cooling water, remove the condenser, and carefully decant the reaction mixture from the boiling chips into an evaporating dish. Avoid breathing the methanol vapors. Place the evaporating dish on the preheated hotplate (70–75° C) in the hood to evaporate all excess methanol.

4. When all of the alcohol has distilled off and only a couple of milliliters of oil remain, remove from the heat, add 25 mL of 5% $NaHCO_3$, and stir until there is no further evolution of CO_2.

5. Separate the small oil layer (the density of methyl salicylate is 1.1738) from the aqueous layer that now contains sodium salicylate. Use a small separatory funnel, available at the storeroom (see Experiment 18 and Figure 18.3 for techniques). Wash the ester with 20 mL of water (in the separatory funnel) to remove remaining sodium salicylate and methanol.

Transfer the oily ester to a vial or a small test tube to be turned in. To be sure, this product is wet and relatively impure, but you can note its distinctive odor and run the following tests:

a. *Test for water solubility:* Add one drop of your methyl salicylate to about 1 mL of water in a test tube. Swirl and tap gently and assess its relative solubility.

b. *Test for acidity:* Run the moist litmus test and the bicarbonate tests on your "wintergreen" as you did on aspirin in part A.

c. *Test for phenolic character:* As was done in part A, test a tiny drop of your "wintergreen" by shaking in a couple of milliliters of water and then adding a drop of ferric chloride solution. Note the color.

Turn in what remains of your aspirin and oil of wintergreen—properly labeled—to your instructor for grading. Clean up and check in condensers, filter flasks, and funnels.

DATA SHEET

Name: ______________________ Lab Section: ______ Date: ______

ESTERIFICATION: ASPIRIN AND OIL OF WINTERGREEN

Data from the Literature

	Formula Weight	mp or bp	Density	Refractive Index	Solubility
Salicylic acid	______	______	xxx	xxx	______
Acetylsalicylic acid	______	______	xxx	xxx	______
Methyl salicylate	______	______	______	______	______

Test Results

	Litmus	$FeCl_3$	$NaHCO_3$
Salicylic acid	______	______	______
Aspirin (crude)	______	______	______
Oil of wintergreen (crude)	______	______	______

Grams of aspirin obtained ______ Theoretical yield: ______

Percent yield ______

Equations for Reactions

Synthesis of aspirin:

Synthesis of oil of wintergreen:

Reaction of aspirin with sodium bicarbonate:

QUESTIONS

1. Aspirin tablets for adults normally contain 5 grains of aspirin—the rest of the tablet being composed of filler and binder material. As little as 10 grams of aspirin in a single dose has been reported to be lethal. If 1 grain equals .0648 gram, how many aspirin tablets might cumulate to this minimum lethal dose?

2. One can purify aspirin by recrystallizing from water, but high temperatures must be avoided to minimize hydrolysis. Thus, one can saturate water at 37° C where the solubility of aspirin is 1 gram per 100 mL. Cooling this saturated solution to 0° C on ice precipitates the aspirin in purer form. If the solubility of aspirin in water at 0° C is 0.2 g/100 mL:

   a. What volume of water should one take to recrystallize 40 g of impure aspirin in accordance with these conditions?

   b. What is the maximum percent recovery you could get by cooling from an initial 37° to 0° C and filtering?

3. Your synthesis of aspirin called for 2 grams of salicylic acid and an excess of acetic anhydride. Suppose that you had started with 1.80 grams of salicylic acid. What would the theoretical yield be?

4. If, after starting as prescribed in question 3, you came out with a pure, dry yield of 1.95 grams of aspirin, what was your percent yield?

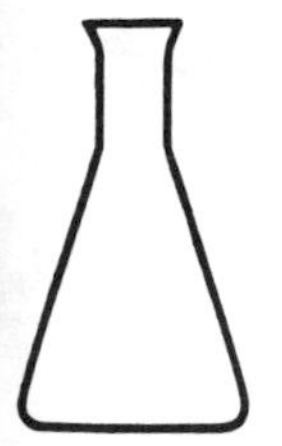

Saponification: The Preparation of Soap

OBJECTIVES

1. To illustrate saponification of a fat.
2. To gain an understanding of the chemical relationships among fats, fatty acids, soaps, and detergents.

PRELABORATORY QUESTIONS

1. What substance is a by-product of soap manufacturing?
2. Explain the difference between fats and oils with regard to physical properties, molecular structures, and sources.
3. What was the function of ashes used in soap making by early civilizations?

DISCUSSION

The ancient Babylonians appear to have used a soaplike material as early as 2800 B.C. This was made by heating fats with ashes. Variations of this same method were used by others throughout history, including ancient Egyptians, Romans, medieval Europeans, and colonial Americans. Today, although the techniques involved in soap making have become highly refined, the fundamental chemical process has remained the same. This process involves hydrolysis of fats to form *glycerol*, a *tri*-alcohol having the structure

$$\begin{matrix} OH & OH & OH \\ | & | & | \\ CH_2- & CH- & CH_2 \end{matrix}$$

, and *soap*, which is a salt of a fatty acid. Ingredients such as perfumes and oils are often added to the soap, and air may be whipped into it to make it float.

A *fat* is the *tri*ester formed from a molecule of glycerol and three fatty acid molecules. *Fatty acids* are long-chain, carboxylic acids such as capric acid (from coconut oil), which has the following structure:

$$CH_3-CH_2-CH_2-CH_2-CH_2-CH_2-CH_2-CH_2-CH_2-\overset{\overset{\displaystyle O}{\|}}{C}-OH$$

Capric acid

The fat formed from glycerol and capric acid is called glyceryl tricaprate and has the following structure:

$$\begin{array}{l} CH_3-(CH_2)_8-\overset{\overset{\displaystyle O}{\|}}{C}-O-CH_2 \\ \qquad\qquad\qquad\qquad\quad\;\; | \\ CH_3-(CH_2)_8-\overset{\overset{\displaystyle O}{\|}}{C}-O-CH \\ \qquad\qquad\qquad\qquad\quad\;\; | \\ CH_3-(CH_2)_8-\overset{\overset{\displaystyle O}{\|}}{C}-O-CH_2 \end{array}$$

Glyceryl tricaprate

A variety of fatty acids is found in plants and animals. Some of these, such as capric, palmitic, and stearic acids, have no carbon–carbon double bonds and are "saturated" fatty acids. Others such as oleic and linoleic acids have one or more double bonds in the hydrocarbon chain and are said to be "unsaturated" fatty acids. A fat may contain both saturated and unsaturated fatty acids in the same molecule.

Some of the common fatty acids and their sources are:

Name	Formula	Source
Butyric	$CH_3(CH_2)_2COOH$	butterfat
Capric	$CH_3(CH_2)_8COOH$	coconut oil
Palmitic	$CH_3(CH_2)_{14}COOH$	palm oil
Stearic	$CH_3(CH_2)_{16}COOH$	animal fats
Oleic	$CH_3(CH_2)_7CH{=}CH(CH_2)_7COOH$	olive oil
Linoleic	$CH_3(CH_2)_4(CH{=}CHCH_2)_2(CH_2)_6COOH$	safflower
Linolenic	$CH_3CH_2(CH{=}CHCH_2)_3(CH_2)_6COOH$	linseed oil

Fat is the term generally applied to *solid* glyceryl triesters, while *oil* is used for the *liquid* triesters. Solid fats come from animal sources (such as lard) and contain a higher percentage of saturated fatty acids than is found in oils. Oils usually originate from plants (corn oil, peanut oil, etc.). Oils are often converted to solids or semisolids (such as margarine) by hydrogenation of some of the double bonds present in the unsaturated fatty acid chains.

The process of hydrolysis of a fat in basic solution is called *saponification*. In ancient times, ashes provided the base necessary for saponification. Saponification is illustrated by the following reaction, which occurs when glyceryl tributyrate is treated with NaOH:

$$
\begin{array}{l}
CH_3{-}(CH_2)_2{-}\overset{\displaystyle O}{\overset{\|}{C}}{-}O{-}CH_2 \\
CH_3{-}(CH_2)_2{-}\overset{\displaystyle O}{\overset{\|}{C}}{-}O{-}CH \\
CH_3{-}(CH_2)_2{-}\overset{\displaystyle O}{\overset{\|}{C}}{-}O{-}CH_2
\end{array}
+ 3\,NaOH \longrightarrow
\begin{array}{l}
HO{-}CH_2 \\
HO{-}CH \\
HO{-}CH_2
\end{array}
+ 3\,CH_3{-}(CH_2)_2{-}\overset{\displaystyle O}{\overset{\|}{C}}{-}O^-\,Na^+
$$

Glyceryl tributyrate Glycerol Sodium butyrate

In this experiment you will carry out a similar reaction, except that you will use a different fat, *glyceryl tristearate*, found in lard. The products are glycerol and the soap, sodium stearate.

MATERIALS NEEDED

Obtain the items of equipment shown in Figure 23.1. These include an evaporating dish, a 400 mL beaker, a ringstand with ring, a wire gauze, and a gas burner. You will also need a stirring rod, a funnel, and filter paper. Chemicals needed for this experiment include lard (or vegetable shortening), 50% NaOH, 0.1 M $CaCl_2$, 0.1 M NaOH, 3 M HCl, saturated NaCl solution, ethyl alcohol, and a liquid detergent.

EXPERIMENTAL PROCEDURE

A. Saponification of Lard

1. **a.** Weigh an evaporating dish on a top-loader or open-beam balance and then weigh into it approximately 10 grams of lard. On the data sheet, describe the lard as to appearance, texture, etc.

   b. Add 10 mL of ethyl alcohol to the lard (this serves as a solvent for the reaction). Mix well with a stirring rod to break up the lard as much as possible. Set up a ringstand with ring, wire gauze, and a 400 mL beaker half full of water, as shown in Figure 23.1. Place the evaporating dish containing the lard on the rim of the beaker and heat the water to boiling while stirring the lard–alcohol mixture.
   CAUTION: Both the lard and the alcohol are *flammable*. Have a wet paper towel nearby so that you can place it over the evaporating dish to extinguish any fire that occurs.

   c. Add 10 mL of 50% NaOH solution to the lard–alcohol mixture. Continue heating and stirring until the material in the evaporating dish becomes solid and no further change is apparent.
   CAUTION: The NaOH is caustic! **Wear goggles** and take care to keep the NaOH off your skin. Any spills should be cleaned up immediately.

   On the data sheet, describe the solidified material resulting from reaction of lard with NaOH.

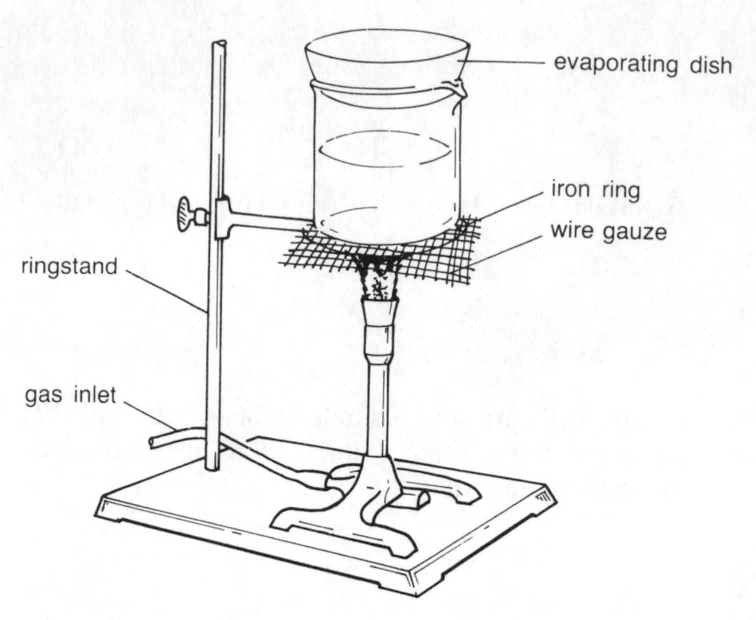

Figure 23.1

2. Add the solidified material from step 1c to 100 mL of saturated NaCl solution in a beaker. Warm gently with stirring, then allow the solution to cool. This process is called "salting out." The high concentration of dissolved NaCl forces the less-soluble soap out of solution. The other product of the reaction, glycerol, is soluble in water. The soap should rise to the surface as a cake. Lift it out (or if it does not coagulate sufficiently, filter the mixture) and wash it with a small amount of water. Use a paper towel to squeeze out the excess water. Describe your soap on the data sheet.
3. Use some of the soap cake to wash your hands. Record your observations on the data sheet.
4. Dissolve *half* of the remaining soap in a minimum amount of *warm* water. Filter the solution.
   - **a.** To 1 or 2 mL of the soap solution in a test tube, add several drops of 0.1 M $CaCl_2$ solution. Record your observations on the data sheet. Hard water contains dissolved Ca^{2+} and Mg^{2+} ions. The fatty acid salts of these ions are insoluble in water, forming "soap scum" or "bathtub rings."
   - **b.** To the remainder of the soap solution add 3 M HCl drop by drop with stirring until no further change takes place. Filter the product from the solution and dry it with a paper towel. The product is stearic acid, $CH_3(CH_2)_{16}COOH$, generated by reaction of hydronium ions from the HCl with sodium stearate. This acid is used to make dripless candles. It burns cleanly, leaves no residue, and does not run down the side of the candle when warm. Its texture is similar to that of the paraffin "waxes." Use the space provided on the data sheet to describe the product (color, texture, etc.).
   - **c.** Attempt to dissolve a small amount of stearic acid in a few milliliters of water. Record your observations on the data sheet.

d. Attempt to dissolve a small amount of stearic acid in a few milliliters of 0.1 M NaOH. Record your observations on the data sheet.

B. Detergents

The following compounds are synthetic detergents:

Sodium lauryl sulfate (a sodium alkylsulfate)

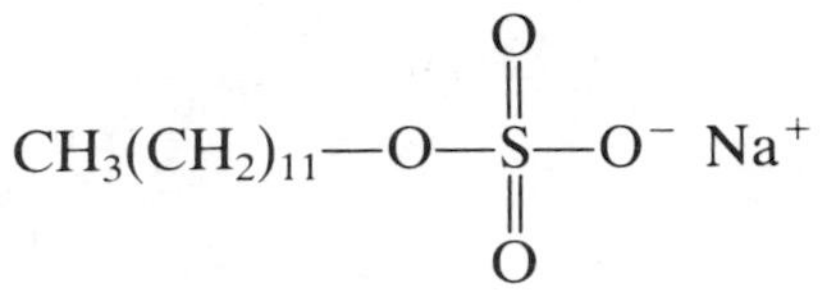

$$CH_3(CH_2)_{11}-O-\underset{\displaystyle O}{\underset{\|}{\overset{\displaystyle O}{\overset{\|}{S}}}}-O^-\ Na^+$$

A sodium alkylsulfonate

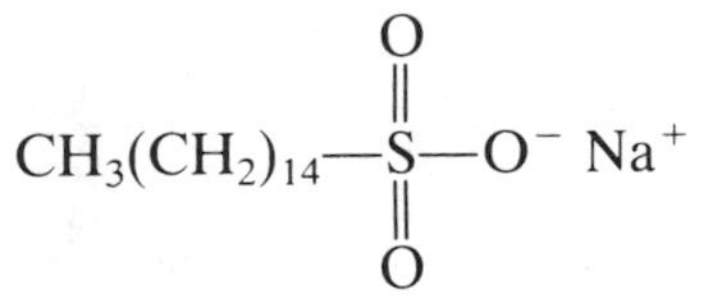

$$CH_3(CH_2)_{14}-\underset{\displaystyle O}{\underset{\|}{\overset{\displaystyle O}{\overset{\|}{S}}}}-O^-\ Na^+$$

A sodium alkylbenzenesulfonate (paraffin base)

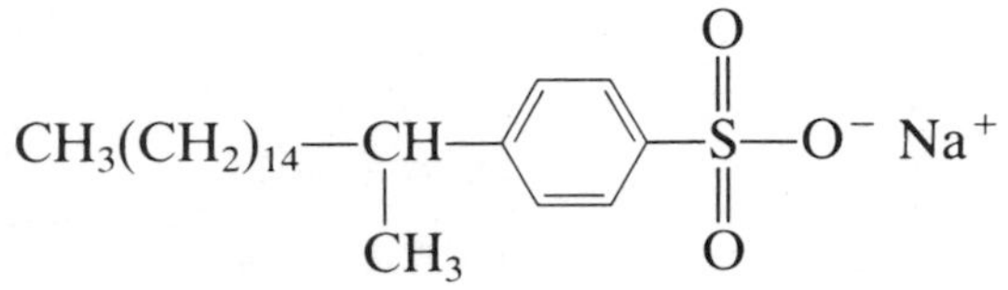

$$CH_3(CH_2)_{14}-\underset{\displaystyle CH_3}{\underset{|}{CH}}-C_6H_4-\underset{\displaystyle O}{\underset{\|}{\overset{\displaystyle O}{\overset{\|}{S}}}}-O^-\ Na^+$$

A sodium alkylbenzene sulfonate (tetrapropylene base)

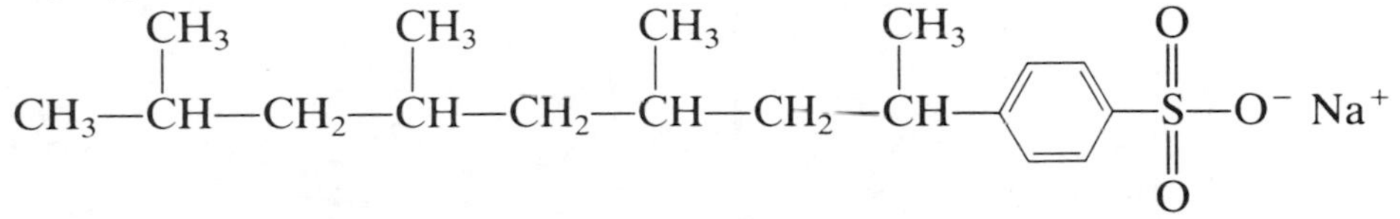

$$CH_3-\overset{\displaystyle CH_3}{\overset{|}{CH}}-CH_2-\overset{\displaystyle CH_3}{\overset{|}{CH}}-CH_2-\overset{\displaystyle CH_3}{\overset{|}{CH}}-CH_2-\overset{\displaystyle CH_3}{\overset{|}{CH}}-C_6H_4-\underset{\displaystyle O}{\underset{\|}{\overset{\displaystyle O}{\overset{\|}{S}}}}-O^-\ Na^+$$

Detergents have the advantage that they are not precipitated by Ca^{2+} and Mg^{2+} ions and therefore may be used in hard water. However, while soaps can be attacked by bacteria and degraded into simpler molecules, some of the early synthetic detergents were not so attacked and passed through sewage treatment plants unchanged, causing foam in rivers and tap water. Modifications in the structures of the hydrocarbon portions of detergent molecules have led to biodegradable detergents (the first three in the above list are biodegradable; the last is not). Many states have declared that nonbiodegradable detergents cannot be sold.

1. On the data sheet, explain how structures of the above detergents differ from those of soaps.
2. Dissolve a few drops of liquid detergent in 10 mL of water in a test tube. Add several drops of 0.1 M $CaCl_2$ solution. Record your observations on the data sheet.

DATA SHEET

Name: ______________________ Lab Section: ______ Date: ____________

SAPONIFICATION: THE PREPARATION OF SOAP

A. Saponification of Lard

1. a. Describe the general appearance and texture of lard.

c. Describe the solid material resulting from reaction of lard with NaOH.

Write a balanced equation (using structures) for the reaction that takes place and name the products.

2. Describe the general appearance of the soap (color, odor, etc.).

3. Does the soap lather? Is it oily?

Compare your soap with a commercial brand you frequently use.

4. **a.** Result of addition of $CaCl_2$ solution:

Why it is difficult to wash with soap in hard water?

Write an equation for the reaction of $CaCl_2$ with sodium stearate.

b. Describe the product resulting from the reaction of sodium stearate with HCl.

Write an equation for the reaction of sodium stearate with HCl.

c. Observations during attempt to dissolve stearic acid in water:

d. Observations during attempt to dissolve stearic acid in 0.1 M NaOH:

Is there a difference in solubility of stearic acid in water compared to the NaOH solution? Why?

B. Detergents

1. How do the molecular structures of detergents differ from those of soaps?

2. Result of addition of $CaCl_2$ solution:

Compare this result with that obtained by adding $CaCl_2$ to the soap solution (part A4).

QUESTIONS

1. How many grams of sodium stearate (formula mass 306) can be made from 10 grams of glyceryl tristearate (formula mass 890)? Refer to your balanced equation on the data sheet, part A1.

2. What factors would you take into consideration in developing a soap or detergent for commercial use?

3. Write the structural formula of glyceryl trioleate.

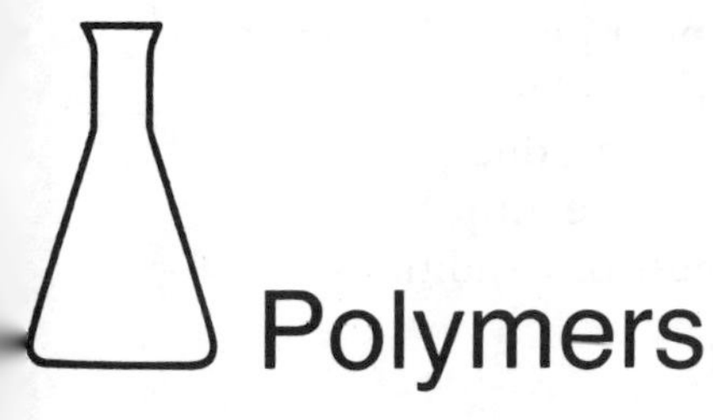

Polymers

OBJECTIVES

1. To gain experience in the synthesis of three types of polymers.
2. To understand the difference between addition and condensation polymerization.
3. To understand the difference between a simple polymer and a copolymer.

PRELABORATORY QUESTIONS

1. How does condensation polymerization differ from addition polymerization?
2. What monomers might be used to synthesize Nylon 64?
3. Benzoyl peroxide has the structure $\emptyset\overset{\overset{\displaystyle O}{\|}}{C}—O—O—\overset{\overset{\displaystyle O}{\|}}{C}\emptyset$. (The symbol Ø is used here to abbreviate $C_6H_5^-$.) It functions as an excellent free radical catalyst by readily breaking down to generate active free radicals and CO_2. Draw electronic structures to illustrate (a) a benzoate free radical and (b) a phenyl free radical.
4. What is a copolymer?

DISCUSSION

This experiment provides an opportunity to make three common polymers. The chemical industry, using sophisticated refinements in procedures, markets enormous quantities of these same polymers. Try your luck as a plastics chemist. You will find the monomers very easy to assemble into polymers.

Polymers are ordinarily mixtures of molecules. That is, the molecules are like a school of fish: Though of the same species, they can and do differ in size. Because they are extremely large, as molecules go, they are hard to dissolve in solvents and they tend to soften or decompose before they actually melt.

The first polymer prepared in this experiment is quite atypical. Hexamethylenetetramine has a definite molecular weight and a high solubility in most polar solvents. It is a *condensation* product of six molecules of formaldehyde and four molecules of ammonia. As a synthetic product, it is probably the most elegant to be synthesized in this course. Its beautifully symmetrical cage structure magically forms under the simplest of conditions:

$$6H_2C{=}O + 4NH_3 \underset{\text{hydrolysis}}{\overset{\text{condensation copolymerization}}{\rightleftharpoons}} (CH_2)_6N_4 \text{ (cage structure)} + 6\,H_2O$$

This reaction is an example of a condensation: Water is split out. By contrast, the next polymer we prepare, polystyrene, is an *addition polymer*—one in which there are no by-products.

Hexamethylenetetramine (also called metheneamine or urotropine) is a white crystalline solid that sublimes, rather than melts, at about 263°C. It burns with a hot smokeless flame and is consequently used by campers in the form of fuel tablets. It is readily soluble in water (note the four 3° amine groups), in which it slowly hydrolizes back to its original monomer components. This feature finds application in its former use as a urinary tract disinfectant. The slow release of formaldehyde is the secret for the disinfectant action. The ammonia, of course, is harmless in low concentration and is excreted as urea and ammonium salts. There are a great many other useful applications of this interesting molecule.

Polystyrene is a true *addition polymer* and is composed of a long chain of monomer styrene units connected in a head-to-tail fashion. It is a clear polymer with excellent insulating properties, a widely used industrial plastic. The overall equation for the polymerization of alkenes like styrene is:

$$\underset{\text{an alkene}}{n\ RCH{=}CH_2} \xrightarrow[\text{addition polymerization}]{\text{catalyst}} \underset{\text{a polyalkene}}{(-\overset{R}{\overset{|}{C}}H-CH_2-)_n}$$

This polymerization of a styrene proceeds via a free radical chain reaction. The catalyst, benzoyl peroxide, produces free radicals (atoms or groups of atoms that possess an unpaired electron). These highly reactive free radicals then initiate the polymerization by activating a monomer molecule, styrene, to a new free radical. This activated monomer then starts a sequence of reactions (propagation) that continues until some event terminates the chain reaction by destroying the free radical nature of the growing active polymer. The following equations summarize the mechanism of this free radical chain reaction:

$$\underset{\text{Benzoyl peroxide}}{\text{Ø}\overset{O}{\overset{\|}{C}}O{:}O\overset{O}{\overset{\|}{C}}\text{Ø}} \xrightarrow{\text{dissociation}} \underset{\text{Benzoate free radical}}{2[\text{Ø}\overset{O}{\overset{\|}{C}}O^{\bullet}]} \xrightarrow{\text{decarboxylation}} \underset{\text{Phenyl free radical}}{2[\text{Ø}^{\bullet}]} + 2CO_2$$

Any of the free radicals can then initiate the chain reaction that follows:

$$[\text{Ø}^{\bullet}] + \underset{\underset{\text{Ø}}{|}}{CH_2{=}CH} \xrightarrow{\text{initiation}} [\text{Ø}CH_2\underset{\underset{\text{Ø}}{|}}{\dot{C}H}] \xrightarrow[\text{propagation}]{(N+1)\ CH_2{=}CH\text{Ø}}$$

monomer (styrene) — active monomer

$$[\text{Ø}CH_2\underset{\underset{\text{Ø}}{|}}{CH}(CH_2\underset{\underset{\text{Ø}}{|}}{CH})_nCH_2\underset{\underset{\text{Ø}}{|}}{\dot{C}H}]$$

active polymer

$$\Big\downarrow [\text{Ø}^{\bullet}]\ \text{termination (coupling)}$$

$$\text{Ø}CH_2\underset{\underset{\text{Ø}}{|}}{CH}(CH_2\underset{\underset{\text{Ø}}{|}}{CH})_nCH_2\underset{\underset{\text{Ø}}{|}}{CH}\text{Ø}$$

polystyrene

Free radicals are extremely reactive and short-lived. Even traces of water destroy them, so when polymerizing styrene with benzoyl peroxide, be sure that everything is *dry*.

The square brackets enclosing each free radical in the equations above are used to signify that free radicals indeed have an exceedingly short and transitory existence.

Nylon is the *copolymer* of *two* different monomers. Amide bonds form with the release of water, so Nylon is a condensation polymer. The most common Nylon is Nylon 66, so called because both monomer units are six-carbon molecules. Nylon 510 is the name given the product of the condensation of pentamethylenediamine (5 C's) and sebacoyl chloride (10 C's):

$$n\ H_2N(CH_2)_5NH_2 + n\ Cl\overset{\overset{O}{\|}}{C}(CH_2)_8\overset{\overset{O}{\|}}{C}Cl \xrightarrow[\text{polymerization}]{\text{condensation}}$$

Pentamethylene diamine — Sebacoyl chloride

$$(-NH(CH_2)_5NH\overset{\overset{O}{\|}}{C}(CH_2)_8\overset{\overset{O}{\|}}{C}-)_n + 2n\ HCl$$

Nylon 510

Proteins are naturally occurring copolymers of monomeric amino acids and have the same recurring amide (peptide) connecting links as Nylon.

The great tensile strength of polyamides like Nylon 66 is due to extensive H-bonding between the linear chains as they align themselves in the continuous dovetailed bundle of the fiber. Segments of just two such aligned polymer molecules, shown below illustrate this secondary bonding. When it is remembered that each H-bond has a strength of approximately 5 kcal per mole, it is not difficult to understand why the cumulative effect of as many as 200 H-bonds per Nylon molecule cumulates to a formidable binding force.

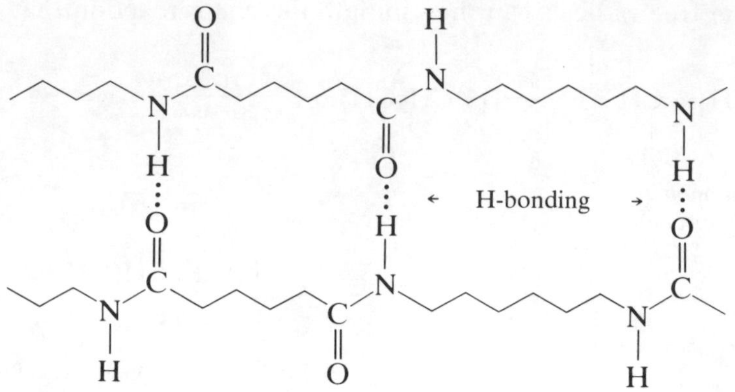

MATERIALS NEEDED

You will need to obtain the following items of equipment for this experiment: a filter flask (250 mL) with one-hole rubber stopper to fit and an appropriate 12 inch length of glass tubing for the making of an air inlet tube, a reflux condenser, and a small Büchner or Hirsch funnel with filter paper to fit. All other materials needed will be provided in the laboratory.

EXPERIMENTAL PROCEDURE

A. Hexamethylenetetramine

Check out a 250 mL suction flask and a reflux condenser. Prepare an air inlet tube by drawing a piece of glass tubing to a pipet-sized tip (see Experiment 1) and set up the evaporation apparatus as illustrated in Figure 24.1. If your lab is equipped with steam cones, you can set this up on the steam heat instead of in the boiling water bath illustrated. The air inlet tube serves to impinge a fine stream of air on the evaporating surface, and at the same time provides enough of a leak to prevent back-up water from the aspirator. Ammonia and formaldehyde fumes are both highly water soluble and are picked up in the aspirator and washed down the drain.

Measure into this evaporator flask 10 mL each of formalin (37% formaldehyde) and concentrated ammonia. Secure the stopper tightly and clamp the apparatus firmly in position. Apply the full force of the aspirator and proceed to heat the water bath (or steam bath) to a gentle boil.

The evaporation of the water in the system is a slow process, so proceed with parts B and C while you wait.

When most of the water has been removed, add another 5 mL of concentrated ammonia and again evaporate as much as possible. When only granular crystals remain, disconnect the aspirator hose *at the aspirator* and fix a Hoffman or a Mohr clamp close to the sidearm to close it off. Fit the flask with a reflux condenser as in Figure 24.2, add 20 mL of absolute alcohol, and heat to a gentle reflux, or as hot as it can be heated in the water bath (or on the steam). After a few minutes of heating, any undissolved material is probably other undesired polymeric forms of the two monomers. Allow any

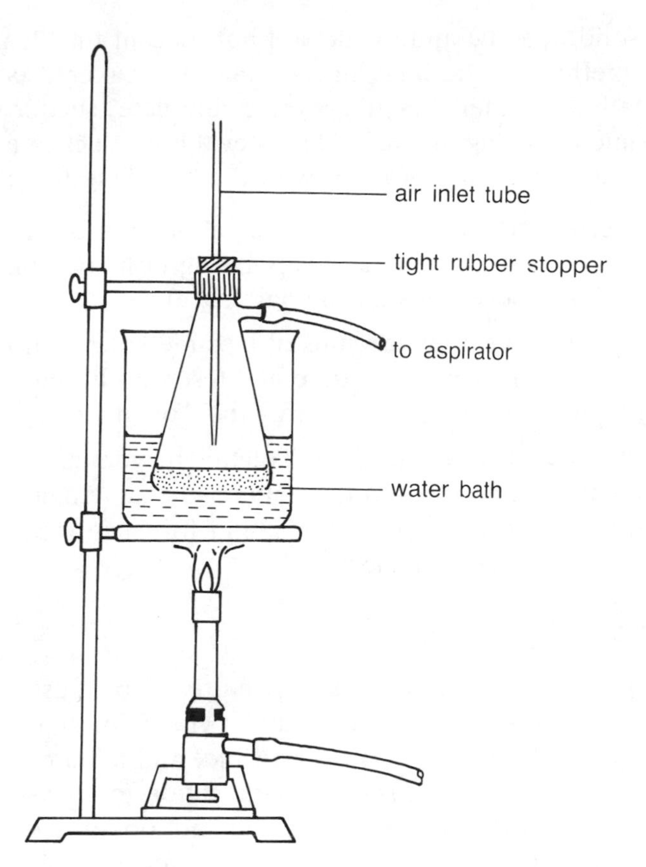

Figure 24.1

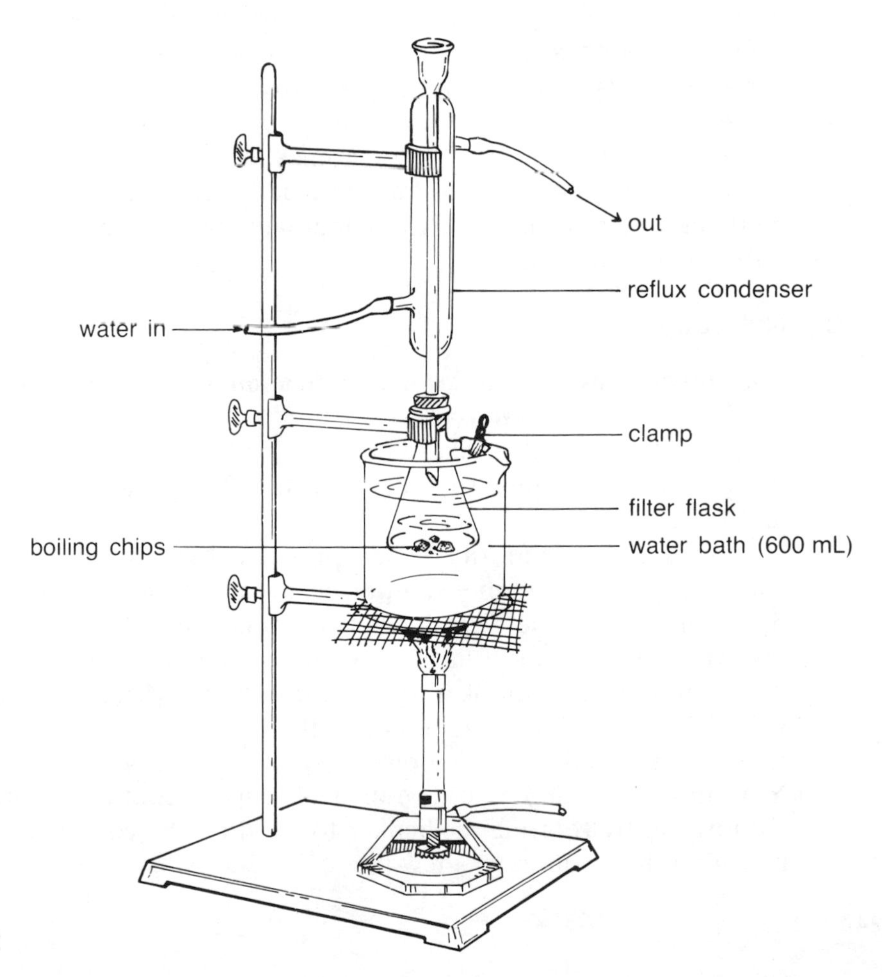

Figure 24.2

remaining solid to settle and, while still hot, decant the clear alcohol solution of the hexamethylenetetramine into a small dry beaker. Cool thoroughly in an ice water bath and, after crystallization is complete, suction filter to dryness (see Experiment 18, Figure 18.1). Transfer the product to a labeled weighing box (see Figure 22.1), and proceed with the following tests:

1. **Ignition characteristics:** Place a small amount of dry hexamethylenetetramine in an evaporating dish and ignite with a match. Record your observations on the data sheet.
2. **Acidity:** Dissolve a few crystals of hexamethylenetetramine in 1 mL of distilled water in a clean test tube and test with litmus paper. Don't throw out the aqueous solution! Save it for the Tollen's test.
3. **Tollen's test:** Prepare 2 mL of Tollen's reagent as described in Experiment 20, part C. Add the aqueous hexamethylenetetramine solution and heat in a water bath. Don't forget to record your observations on the data sheet.

B. Polystyrene

This is a good place to get rid of a chipped test tube. Just be sure it is clean and dry. In it mix 1–2 mL of styrene and a crystal of benzoyl peroxide. Heat in a boiling water bath, in a hood if possible, and monitor the polymerization progress from time to time. Usually solidification to a glassy solid is complete within an hour. If not, you can set it aside and look at it during the next lab. If it is then still sticky, you probably used too much catalyst and initiated too many active monomers, and as a consequence they were unable to grow to a sufficient length to form a solid polymer. If it has hardened completely, you can break away the glass tube to recover your polystyrene.

Another clear glasslike polymer, lucite (Plexiglas), can be prepared in the same way by substituting the monomer methyl methacrylate for the styrene. If you try this, however, use a water bath heated to just 70°C since methyl methacrylate boils at 100°C. Polymethylmethacrylate is an especially fine imbedding material.

C. Nylon 66

Both monomers used in this polymerization are noxious. Work in the hood if possible and avoid skin contact!

In a clean 50 mL beaker pour about 10 mL of the 5% aqueous hexamethylenediamine solution provided. To this add 1 mL of 10% NaOH and mix.

In a second clean, dry beaker procure about 10 mL of the 5% adipyl chloride–petroleum ether solution. Carefully pour the adipyl chloride solution down the side of the tilted beaker of amine solution so that it layers on the top with a minimum of mixing. A film of Nylon forms at the interface.

Fashion a small hook at the end of a 6 inch piece of copper or aluminum wire. Use this to snag the center of the interfacial Nylon film. Avoid agitating the mixture. Pull the Nylon out slowly and steadily with a twisting motion. You may use a large test tube or beaker as a spool and continue the withdrawal by rotating the beaker to wind up the continuously forming Nylon thread. When you have a few feet of thread, wash it thoroughly under the tap and pour the monomer solutions into the container labeled WASTE MONOMERS. Do not pour into the sink! Not only are the monomers lachrymators with foul odors, polymerization in the drains is an almost irreversible process.

Clean up all glassware and show your products to the instructor for approval.

DATA SHEET

Name: ______________________________ Lab Section: ______ Date: ____________

POLYMERS

Instructor's approval:

____________________	____________________	____________________
Hexamethylenetetramine	Polystyrene	Nylon

Results of tests on hexamethylenetetramine:

____________________	____________________	____________________
Ignition	Litmus	Tollen's test

Equation for the synthesis of polystyrene:

Equation for the synthesis of Nylon 66:

QUESTIONS

1. Lucite is a condensation polymer of methyl methacrylate,

$$CH_2{=}\underset{|}{\overset{CH_3}{C}}{-}CO_2CH_3$$

Write an equation for the polymerization of methyl methacrylate.

2. Hexamethylenetetramine (urotropine) is a large, nonionic molecule with a molecular weight of 140. From an inspection of its structure at the right advance an explanation for:

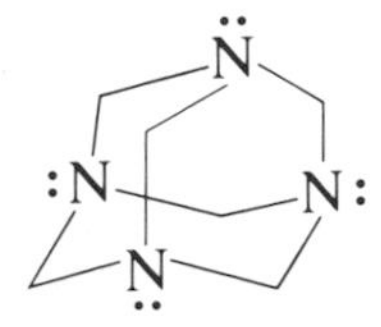

a. Its basicity:

b. Its unusual water solubility (1 g in 1.5 mL):

3. In the commercial preparation of Nylon, many tiny fibers of the polymer are spun, twisted, and stretched into larger threads in a vacuum chamber, and the tensile strength of the Nylon is greatly improved in the process. (Recall that our laboratory threads were not particularly strong.) Explain, at the molecular level, how the spinning, twisting, stretching, and drying can contribute to the tensile strength of the Nylon.

4. When subjected to combustion, the nitrogen in urotropine (hexamethylenetetramine) is oxidized to N_2 gas. (The carbon and hydrogen form carbon dioxide and water, of course.) Using the condensed molecular formula $(CH_2)_6N_4$, write a balanced equation for the combustion that takes place when hexamethylenetetramine is used as an emergency fuel.

5. Urotropine is soluble enough to negotiate the digestive tract and reach the urinary tract, where it slowly hydrolyzes back to its monomeric NH_3 and HCHO. The released formaldehyde is believed to be the active antiseptic, denaturing the protein material of infecting bacteria. Why couldn't one simply take a dilute formaldehyde solution orally and accomplish the same thing?

EXPERIMENT 25

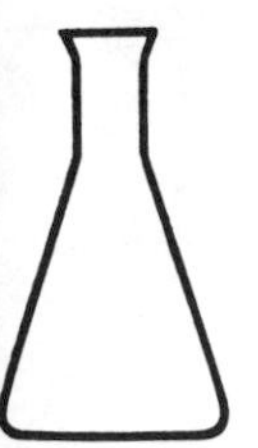

Optical Isomerism: A Study with Molecular Models

OBJECTIVES

1. To learn how to detect planes and centers of symmetry in molecular models and thereby to establish the chirality or achirality of organic compounds.
2. To gain an understanding of the terminology of stereochemistry.
3. To improve your ability to visualize the three-dimensional character of a molecule from its molecular formula.

PRELABORATORY QUESTIONS

1. What are enantiomers?
2. What are diastereomers?
3. What is meant by chirality?
4. The elements of symmetry include planes, centers, and rotation axes of symmetry. In this experiment you will identify planes and centers of symmetry. Define (a) plane of symmetry and (b) center of symmetry.
5. What is a meso compound?
6. What is a racemic mixture?

DISCUSSION

In Experiment 14 isomerism and molecular structure were introduced. Covered there was the simplest type of stereoisomerism: *cis* and *trans*, or geometric isomerism. In this experiment, the relatively subtle differences in another type of stereoisomer, the optical isomer, will be demonstrated. Since a clear understanding of the phenomenon is contingent upon the special stereochemical vocabulary used to describe it, it is strongly recommended that you study the following section on theory as well as your textbook exposition of stereochemistry before proceeding with the laboratory assignment.

Stereoisomers are different structurally only in the spatial arrangement of their atoms or groups. Basically, there are just two kinds of stereoisomers: enantiomers and diastereomers.

Enantiomers are by definition nonsuperposable mirror images of one another. They are related to each other as are your right and left hands. They are said to be *chiral* or to possess chirality (handedness). Individually, they have none of the elements of symmetry such as planes, centers, or axes of symmetry.

All enantiomers have identical physical and chemical properties except for two. Physically, they have equal but opposite rotational effects on the passage of a beam of plane-polarized light. One enantiomer, the *dextro isomer*, rotates the plane of light to the right (clockwise). Its *optical isomer*, the *levo isomer*, will rotate the same plane-polarized light equally, but to the left (counterclockwise). This *specific rotation* exhibited by enantiomers *is the one physical handle* the chemist has for monitoring the presence and purity of chiral compounds. When a 50–50 mixture of dextro and levo enantiomers is subjected to a beam of plane-polarized light, the optical rotation is reduced to zero. Such a mixture is called a racemate—or *racemic mixture*. When an optically active enantiomer is capable of switching from its original configuration to the configuration of its mirror image, the phenomenon is called *racemization*.

Diastereomers are stereoisomers that may or may not be chiral. They are simply not mirror images of one another. If indeed each of a pair of diastereomers is chiral and exhibits optical activity, the rotations are never of equal and opposite degree as is the case for a pair of enantiomers. Diastereomers *do* have different chemical and physical properties and therefore can be separated by the usual chemical and physical means. Actually, then, diastereomers include all stereoisomers that are not enantiomers. The single necessary criterion for establishing chirality is the existence of the nonsuperposable mirror image relationship. If this can be shown, the stereoisomers are enantiomers. If not, they must be diastereomers. Simple pairs of diastereomers are illustrated by the achiral *cis* and *trans* isomers studied in Experiment 14.

A *chiral atom* can be spotted in a structural formula. If an atom has four *different* atoms or groups covalently bonded to it, that atom is chiral. It is lacking in all of the elements of symmetry and is therefore chiral. Usually, the presence of chiral atoms causes the molecule to be chiral also. In certain cases, however, a chiral atom may be compensated for in the molecule by another chiral atom that is its mirror image. If a plane of symmetry can be constructed between two such chiral atoms, the compound is called a *meso isomer*, an optically inactive stereoisomer by virtue of internal compensation. The one chiral atom cancels the rotation of its mirror image atom.

Why do we make such a fuss over the subtle and minor difference between enantiomers? It is simply because chiral molecules are fundamental to life itself. Amino acids, enzymes, hormones, sugars, and a continually growing list of physiologically important molecules are chiral. What's more important, in almost every case, just one of the enantiomeric forms is physiologically active. One cannot, for example, substitute dextro-epinephrine for the naturally occurring levo-epinephrine and get the expected cardiac stimulation. Chiral molecules react with other chiral molecules stereospecifically. That is to say, one enantiomer will react with another chiral molecule at quite a different rate than will its mirror image. This *stereospecificity* in reaction rates is the *one chemical difference* between enantiomers. An enzyme or hormone is specific simply because it is chiral and will physically and chemically "fit" only one specific chiral substrate. To study physiology, one must ultimately reduce a process to its chiral, molecular level; molecular biology is basically applied stereochemistry.

MATERIALS NEEDED

A box of ball and stick molecular models.

EXPERIMENTAL PROCEDURE

Select a laboratory partner, and each of you check out a box of molecular models. The color code and differences in bonding are the same as those illustrated in Experiment 14. Proceed through the experiment and answer the questions on the data sheet pertinent to each step taken. When in doubt, ask your lab instructor for help.

A. Chlorobromoiodomethane, ClBrICH

Build a model—each of you—of this compound. Make one the mirror image of the other. Answer questions 1 and 2 on the data sheet.

Exchange any two atoms (e.g., Cl and I) on *one* of the models. Answer questions 3, 4, and 5.

B. Glycine, $H_2NCH_2\overset{\overset{\displaystyle O}{\|}}{C}—OH$

Construct a model of glycine, along with a model of its mirror image. Study their elements of symmetry and answer questions 6 through 8. Remember that there is free rotation about single bonds and that you can rotate groups attached by single bonds to any conformation desired. If the group on one molecule can be rotated, then obviously the same group on its mirror image possesses the same rotational freedom.

C. The Chlorination of 2-Butene; $CH_3CH{=}CHCH_3 + Cl_2 \rightarrow CH_3\underset{\underset{\displaystyle Cl}{|}}{C}H—\underset{\underset{\displaystyle Cl}{|}}{C}HCH_3$

Build a model of *cis*-2-butene as was done in Experiment 14. Examine it for elements of symmetry and answer questions 9 through 12.

Without dismantling the *cis* model, prepare its *trans* isomer. You now have a stereoisomeric pair. Answer questions 13 through 16.

Build a second *cis*- and a second *trans*-2-butene. You will use the four models to pick your way through a hypothetical stereospecific reaction. We must classify this as a hypothetical reaction because we shall assume that chlorine adds to an alkene by *cis*-addition. In actuality, Cl_2 undergoes a *trans*-addition (also stereospecific), but our models do not lend themselves to the configuration inversion that is a part of *trans*-addition. Simply assume, for this experiment, that Cl_2 does indeed undergo *cis*-addition, as do a great number of other reagents. We can then demonstrate what is meant by a *stereospecific reaction:* a reaction in which stereochemically different reactants give stereochemically different products.

One of you will work with the pair of *cis* models, the other partner with the *trans* pair. Examine the diastereomer you are working with. Note that if one of the two bonds of the double bond is broken, it is as though the reactant that broke the bond attacked it from a single side. A *cis* addition is the result of attaching chemical groups (in our experiment, chlorines) to the points of severance of the bond that was broken. In other words, both addends come in on the same side when a *cis* (syn) addition is accomplished. You can see, of course, that the other side of the molecule (other bond) must be equally vulnerable to chemical attack.

From the two identical models you are working with (*cis or trans*), disconnect the front-side spring on the one and the back-side spring on the other, and insert pegs in the holes from which the springs were removed. Attach chlorines to these vacant bonds. From a statistical point of view, your two product molecules must represent the composition of the product of a *cis* chlorination of *cis*- (or *trans*-)2-butene. Examine your two product molecules carefully for elements of symmetry (plane or center). One of you will get a single *meso* isomer. The other of you will get a racemate—a pair of nonsuperposable mirror images. Answer questions 17 and 18.

Study the meso compound and answer questions 19, 20, and 21.

You have, then, simulated a stereospecific synthesis to prepare a total of three stereoisomers: a pair of enantiomers, and a meso compound that is itself the diastereomer of either of the enantiomers.

D. The Amino Acid Threonine, $CH_3CH(OH)—CH(NH_2)CO_2H$

One can predict the total possible number of stereoisomers in most molecules by applying *van't Hoff's rule*. This rule states that *the total possible number of stereoisomers for a given molecular structure is equal to 2^n, where n is the number of different* chiral atoms in the molecule. Answer questions 22, 23, and 24.

Use models if necessary to draw the absolute configurations of the structures called for in questions 25 and 26.

E. The Dichlorocyclobutanes

There are five dichlorocyclobutanes, and one of these is chiral, so if all stereoisomeric possibilities are counted, there are six of them. Use models to assist you in answering questions 27 through 33.

When you have finished, disassemble the models, put the boxes back in order, and check them in.

DATA SHEET

Name: ______________________ Lab Section: ______ Date: ________

OPTICAL ISOMERISM: A STUDY WITH MOLECULAR MODELS

I. Using the tetrahedron convention at the right, draw the mirror image of the structure illustrated.

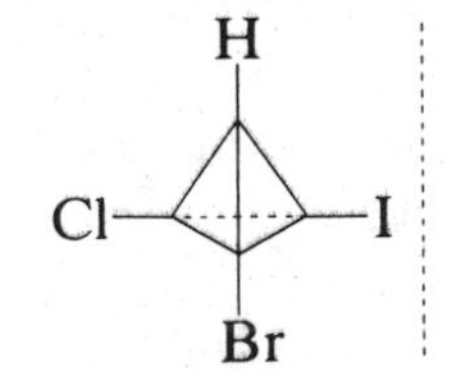

1. Are the mirror images of chlorobromoiodomethane superimposable? ______________

2. Do either of the chlorobromoiodomethanes constructed have a plane of symmetry? ______________

3. Are the two models now mirror images? ______________

4. Are the two models now superposable? ______________

5. If one mole of each of the modeled species were mixed, would the mixture exhibit optical activity? ______________

II. Using the ball and stick convention at the right, illustrate the mirror image of the glycine molecule as it is illustrated.

CO_2H

H NH_2

H

6. Are the mirror image models of glycine superposable? ______________

7. Does either model of glycine have a plane of symmetry? ______________

8. Which model(s) of glycine is/are chiral? ______________

III. Draw the structures of *cis*- and of *trans*-2-butene:

9. Does *cis*-2-butene possess any chiral carbons? ______________

10. Does *cis*-2-butene possess a plane of symmetry? ____________

11. Does *cis*-2-butene possess a center of symmetry? ____________

12. Is *cis*-2-butene a chiral molecule? ____________

13. Does *trans*-2-butene have a plane of symmetry? ____________

14. Does *trans*-2-butene have a center of symmetry? ____________

15. Are *cis*- and *trans*-2-butene superposable? ____________

16. What kind of stereoisomers are *cis* and *trans*-2-butenes? ____________

17. The pair of enantiomers produced from one of the *cis–trans* isomers is:

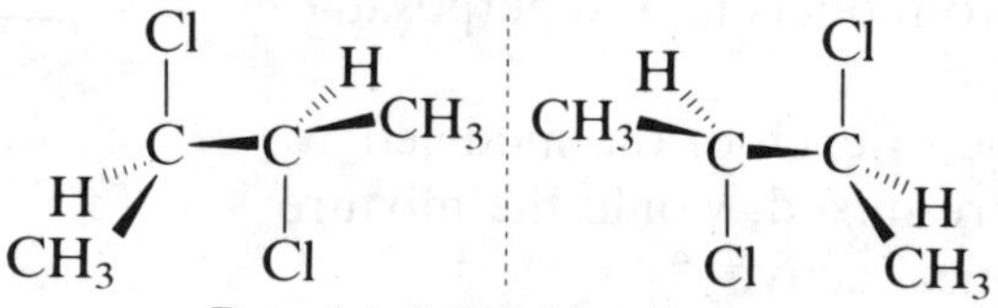

D and L-2,3-dichlorobutane

Did the *cis*- or the *trans*-2-butene produce the pair of enantiomers? ____________

18. Using the convention of question 17, draw at the right the structure for *meso*-2,3-dichlorobutane.

19. How many chiral atoms are there in the *meso*-2,3-dichlorobutane? ____________

20. Is *meso*-2,3-dichlorobutane optically active? ____________

21. What element of symmetry does *meso*-2,3-dichlorobutane possess? ____________

IV. **22.** How many chiral atoms are there in threonine? ____________

23. From van't Hoff's rule, how many stereomers having the molecular structure of threonine are possible? ____________________

24. Can threonine have a *meso* form? ____________________

25. The ball and stick representation of the absolute configuration of the physiologically active L (−)-threonine is shown at the right. Using this convention, draw the structure for D (+)-threonine.

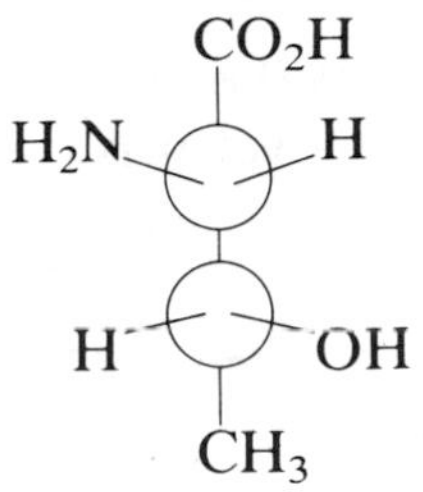

CO_2H
H_2N — — H
H — — OH
CH_3

D (+)-threonine

D- and L-allothreonine

26. In the space indicated above to the right, draw the correct configurations for the remaining two stereomers (stereoisomers) of threonine. These are called D- and L-allothreonine, but you need not attempt to designate which is D and which is L. Use the same convention as you used to draw D (+)-threonine.

V. One of the six isomeric dichlorocyclobutanes is shown at the right and is called *cis*-1,2-dichlorocyclobutane.

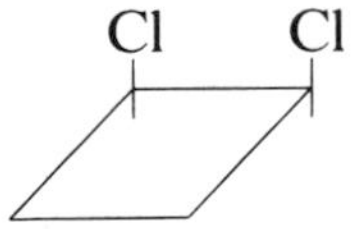

Cl Cl

27–31. Using the convention illustrated, draw structures for the *remaining* five dichlorocyclobutanes and name each of them.

32. Draw circles around the isomeric dichlorocyclobutanes above that are chiral and therefore optically active.

33. Note that the *cis*-1,2-dichlorocyclobutane illustrated above has two chiral carbons—yet it is optically inactive. Why?

EXPERIMENT 26

The Physical and Chemical Properties of Carbohydrates: The Identification of an Unknown

OBJECTIVES

1. To learn, and to be able to test for, the fundamental differences in physical and chemical properties of the common carbohydrates.
2. To be able to identify any of the nine common carbohydrates from the results of tests used in this experiment.
3. To understand the primary structural differences between the nine common carbohydrates.

PRELABORATORY QUESTIONS

1. Draw the open chain and the hemiacetal structures for D (+) = glucose. (Draw α−D(+)-glucopyranose for the hemiacetal representation.)
2. What is a "reducing sugar"?
3. What happens to a sugar while it undergoes mutarotation?
4. How does an aldose differ from a ketose?
5. All monosaccharides are highly water soluble. On the basis of their structure, why is this completely predictable?

DISCUSSION

It is said that once the chemistry of glucose is mastered, the rest of carbohydrate chemistry "comes naturally." This experiment will demonstrate the similarities and the subtle differences existing between glucose and other carbohydrates. Nine of the most common carbohydrates have been selected for the study, and some of their physical and chemical characteristics are tabulated in Table 26.1. You will be issued one of these nine and charged

Table 26.1 Properties of Nine Common Carbohydrates

Property or Test	Glucose	Fructose	Galactose	Sucrose	Maltose	Lactose	Starch	Dextrin	Cellulose
Water solubility	+	+	+	+	+	+	sparing	sparing	–
Ether solubility	–	–	–	–	–	–	–	–	–
Charring (w/concd H_2SO_4)	+	+	+	+	+	+	+	+	+
Melting point	80–90d	103–5d	118–20d	160–186d	100–20d	200–20d	d	d	d
Fermentation (w/yeast)	+	+	–	+	+	–	+	+	–
Benedict's test	+	+	+	–	+	+	–	–	–
Seliwanoff test (<30 sec)	–	+	–	–	–	–	–	–	–
Iodine test	–	–	–	–	–	–	blue	sometimes red	–
Taste (sweetness scale)	0.74	1.5	0.55	1.0	0.33	0.2	0.0	0.0	0.0
Osazone, mp and time	216–17°	216–17°	201°	216–17°	206°	200°d	–	—	—
for ppt appearance	4–5 min	2 min	15 min	30 min		>2 hrs	(unless heated till hydrolyzed)		
Specific rotation (after mutarotation)	+52.7°	−92.0°	+80.2°	+66.5°	+130.4°	+52.3°	–	—	—

with its identification. Plan your own analysis using any or all of the tests described, and any additional devices suggested in the literature. Solutions (or suspensions) of the nine carbohydrates listed in Table 26.1 will be provided for you to run comparison tests.

MATERIALS NEEDED

Check out a fermentation tube, a melting point bath, and an unknown carbohydrate.

EXPERIMENTAL PROCEDURE

Prepare a 2% solution of your unknown—if it is water soluble—by dissolving about 2 g in 100 mL of distilled water. You will use this solution for tests 1, 2, 3, 4, 7, and 9. Record the solubility characteristics observed.

It will be somewhat difficult to identify unconditionally and absolutely a couple of the unknowns. However, in these cases the identity is easily reduced to one of two possibilities. Should you find yourself in this dilemma, simply report your first and second choices and list your reasons for this speculation. Check the literature in such an event!

A. Fermentation

Set this test up first as it is the most time consuming, and the fermentation can then proceed while you do other tests.

The common hexoses, with the exception of galactose, are fermented by yeast according to the equation:

$$C_6H_{12}O_6 \xrightarrow{\text{zymase}} 2CH_3CH_2OH + 2CO_2\uparrow$$

The presence of other hydrolytic enzymes in yeast (amylase, sucrase, and maltase) causes all of the common disaccharides except lactose to hydrolyze to fermentable monosaccharides, so yeast also acts on *them.* Although we do not provide pure isolated enzymes for this experiment, it should be apparent that the selective action of such specific enzymes as sucrase, maltase, lactase, and emulsin would provide excellent analytical options.

In either a fermentation tube (Figure 26.1d) or a large test tube (8 inch) place 4 mL of yeast suspension and about 30–35 mL of the carbohydrate solution to be fermented. Mix well. If you do not have a fermentation tube, insert a small (4 inch) test tube upside down in the large test tube (Figure 26.1a). Place the palm of your hand over the mouth of the large tube and invert so that the smaller test tube is filled (Figure 26.1b). Return the large test tube to the upright position. The smaller tube should now be completely filled and resting upside down within the solution in the large tube (Figure 26.1c). Set the assembly in a water bath maintained at 37°C and see if bubbles of CO_2 collect within the smaller tube. If none have appeared within 2 hours, it can be assumed that your solution is *not* acted upon by yeast. To confirm the fact that collected gas bubbles are indeed CO_2, introduce with a bent tip

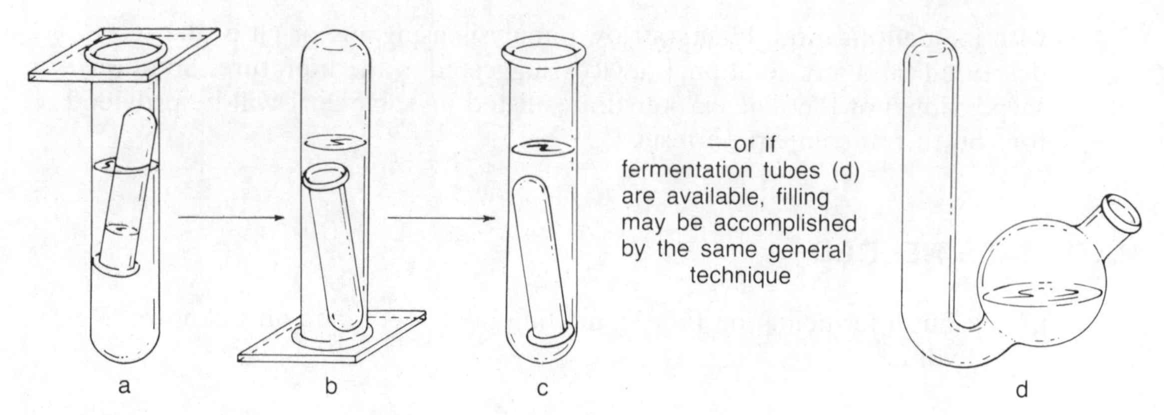

Figure 26.1

dropping pipet a few milliliters of concentrated NaOH at the bottom of the small tube. Carbon dioxide dissolves in strong base solutions. Write the equation for this reaction while you wait for the NaOH to diffuse to the gas. Record your observations.

B. Benedict's Test for Reducing Sugars

Any aldose or ketose having a potentially free carbonyl group will reduce Benedict's solution. A hemiacetal or hemiketal has a potentially free carbonyl group because it is at all times in dynamic equilibrium with its aldehyde or keto precursor. Glycosides are not as readily hydrolyzed to their hemiacetal units and therefore resist oxidation by Benedict's reagent. Thus sucrose, starch, dextrins, and cellulose are not reducing carbohydrates. All the other common carbohydrates included in this experiment are reducing sugars. See Table 26.1.

Ketoses are more extensively oxidized than are the aldoses and generally reduce about four times as much Benedict's reagent per mole. Fehling's solution and Tollen's solution can be used also to determine reducing ability in sugars. Try them if time permits. (See Experiment 20 for a description of Tollen's silver mirror test.)

Place 2 mL of Benedict's reagent in a small test tube and heat in a gently boiling water bath for a couple of minutes. Add five drops of the solution to be tested, *mix thoroughly,* and allow it to heat 5 minutes longer in the bath. A color change from blue (Cu^{2+}) through green ($Cu^{2+} + Cu_2(OH)_2$) to yellow ($Cu_2(OH)_2$) and orange ($Cu_2(OH)_2 + Cu_2O$) to a brick red (Cu_2O) indicates that the carbohydrate being tested is a reducing sugar. The time taken for the appearance of the colors is significant.

Fructose, for example, reacts faster than glucose or galactose. Since polysaccharides will hydrolize in water when extensively heated, even they will eventually bring about some reduction of the Cu^{2+}. The reaction with an aldose can be generalized:

$$RC(=O)—H + 2Cu(OH)_2 \xrightarrow{\Delta} RC(=O)—OH + Cu_2O\downarrow + H_2O$$

Record your observations on the data sheet.

C. The Seliwanoff Test for Ketoses

When a ketose is heated with a strong mineral acid, hydroxymethyl furfural is formed. This compound forms a red complex with resorcinol (1,3-dihydroxybenzene). Seliwanoff's reagent is a freshly prepared solution of resorcinol in concentrated HCl and *when freshly prepared* will give a red color rapidly with ketoses and more slowly with aldoses. Thus the timing must be carefully observed in order that the test distinguish between aldohexoses and ketohexoses. Check against knowns provided in the lab.

Prepare the Seliwanoff reagent just before using by adding 2 mL of 0.5% resorcinol solution to 7 mL of concentrated HCl and diluting to 20 mL with water. Mix well and use as soon as possible. Heat 5 mL of this Seliwanoff reagent to boiling, add ten drops of the carbohydrate solution, quickly mix, and boil for less than 20 seconds. A red color at this time is indicative of a ketohexose. (A blue to green color is characteristic of a pentose.) If the red color develops after 20 seconds, it is likely that the unknown is an aldohexose. Record your observations on the data sheet.

D. The Iodine Test for Starch

Starch forms a dark blue complex when treated with iodine. Erythrodextrin and glycogen give red iodine complexes. Amylodextrins give a blue and achrodextrins and cellulose no color at all. These colored complexes become colorless when heated, but the color comes back again upon cooling. A single drop of I_2/KI solution is enough to give an unmistakable test on a solution or a solid. Iodine stains on starched hospital uniforms range from reds to blue. They may disappear under a hot iron but return on cooling.

Test a milliliter of the solution of your unknown by adding one drop of I_2/KI reagent. Record your observations on the data sheet.

E. Dehydration

Since carbohydrates are polyalcohols, it is to be expected that they will readily dehydrate when heated or when treated with concentrated H_2SO_4. Thus the cellulose in paper can be partially dehydrated with concentrated H_2SO_4 to yield parchment, and sucrose can be partially dehydrated to caramel by heating. Any dry powdered carbohydrate will dehydrate to a spongy mass of carbon if stirred for a time with concentrated sulfuric acid. Try it on a portion of your solid unknown and verify that it is indeed a carbohydrate. Record your observations on the data sheet.

F. Taste

The sweetest sugar is fructose, which is followed by sucrose and glucose in that order. Of the common disaccharides, lactose is the least sweet. Dextrins may have a detectable sweetness if small enough in size, but starch and cellulose are not sweet, unless first digested. If you chew a piece of white bread long enough, it will begin to taste sweet due to the action of the ptyalin in the saliva. Of course, it is never advisable to taste laboratory chemicals, but try it at home some time.

G. Osazone Formation

Carbohydrates with free or potentially free carbonyl groups react with phenylhydrazine to form osazones. The number one and two carbons of the aldose or ketose undergo condensation-oxidation-condensation to form nicely crystalline, colored derivatives according to the overall equations:

$$\underset{\text{an aldose}}{\begin{array}{l}\text{CHO}\\ |\\ \text{CHOH}\\ |\end{array}} \xrightarrow{3\phi\text{NHNH}_2} \underset{\text{an osazone}}{\begin{array}{l}\text{CH=NNH}\phi\\ |\\ \text{C=NNH}\phi\\ |\end{array}} \xleftarrow{3\phi\text{NHNH}_2} \underset{\text{a ketose}}{\begin{array}{l}\text{CH}_2\text{OH}\\ |\\ \text{C=O}\\ |\end{array}}$$

Prepare a phenylhydrazine reagent by mixing 10 mL of water, 3 mL of glacial acetic acid, and 3 mL of phenylhydrazine. **Avoid getting phenylhydrazine on your hands—it is absorbed in the skin and is poisonous!** Add 2 mL of this reagent to 5 mL of the sugar solution to be tested, mix, and heat in a water bath at 100° for up to 30 minutes. Note the time required for the formation of crystals. If the procedure is followed carefully, the times of formation listed in Table 26.1 are fairly duplicable. As you can see from the table, the melting points of the osazones are not especially helpful for this particular identification, but you may wish to determine them anyhow. Simply suction filter, dry, and determine the melting (or decomposition) point.

H. Solubility

Like almost all polyfunctional organic compounds, the carbohydrates are water soluble (unless very large) and ether insoluble. There is no point in testing for ether solubility in this experiment since no carbohydrates are ether soluble. Carbohydrates are, however, easily distinguished from lipids by solubility tests. Lipids are ether soluble and water insoluble. The mono- and disaccharides have sparing solubilities in alcohols. See Table 26.1. This characteristic is described for your information only. You established the water solubility of your unknown when you prepared (or attempted to prepare) its solution at the beginning of the experiment.

I. Optical Rotation

Being replete with chiral carbons, the carbohydrates have measurable and characteristic specific rotations—when they are soluble. If a polarimeter is available and your unknown is water soluble, dissolve about 1 g (exactly weighed) in exactly 10 mL of distilled water (use a 10 mL volumetric flask), and mix thoroughly. Ask your instructor to show you how to use the polarimeter and determine the observed rotation. The specific rotation is then calculated from the formula:

$$[\alpha]_{\mathrm{D}}^{\mathrm{T}} = \frac{\text{observed}}{l \times d}, \text{ where } l \text{ is the length of the polarimeter tube in dm}$$
$$d \text{ is } \mathrm{g_{cpd/mL}} \text{ of the solution}$$

The carbohydrates issued as unknowns are probably already equilibrium mixtures of α and β forms, so you need not attempt to observe the phenomenon of mutarotation.

Specific rotations are listed in Table 26.1.

J. Decomposition Points

As you might expect, the sugars, being all polyalcohols, dehydrate easily when heated. When melting points are determined one usually observes first a shrinking and colorization and then an evolution of gas (water vapor) seen as a foaming as decomposition begins. Carbohydrates simply caramelize as they are heated. Decomposition points are uncertain; they depend on the nature and rate of heating and are seldom useful for identifying specific compounds. They do, however, tell a great deal about the stability of a compound, and so help greatly in establishing the class to which the compound belongs. For the purposes of this experiment, however, determining the decomposition point of your unknown is quite unnecessary. You already know that it is a carbohydrate.

DATA SHEET

Name: ______________________ Lab Section: ______ Date: ____________

THE PHYSICAL AND CHEMICAL PROPERTIES OF CARBOHYDRATES: THE IDENTIFICATION OF AN UNKNOWN

Unknown number: ____________ Unknown is: ____________

Tests

Water solubility: ____________ Seliwanoff test: ____________

Decomposition point: ____________ Iodine test: ____________

Fermentation: ____________ Dehydration test: ____________

Benedict's test: ____________ Osazone (mp): ____________

Specific rotation = ____________

Other observations and tests:

QUESTIONS

1. A certain carbohydrate is not soluble in water and gives a red complex when treated with an iodine solution. What does this tell you about the probable identity of the compound?

2. One of the carbohydrates listed in Table 26.1 was found to be water soluble. It decomposed, charred, and burned completely away when heated slowly to ignition on a porcelain spatula. It did not ferment with yeast at 37°. Its optical rotation was observed to be + 2.8° when a solution of 5.304 g in 100 mL of solution was examined in a polarimeter using a 1 dm polarimeter tube. What is the identity of this carbohydrate?

3. Another sugar from the list in Table 26.1 gave a positive Benedict's test, fermented with yeast, formed an osazone in less than 5 minutes, and gave an immediate red color when heated with Seliwanoff's reagent. What is its possible identity?

4. It is easy to see that the chemical evidence for the conclusion drawn in question 3 is not unequivocal. Given this situation, what would be the best way by means of a lab check to confirm (or reject) the conclusion drawn?

5. When freshly prepared, α-D-glucopyranose has a specific rotation of + 112.2°. Freshly prepared β-D-glucopyranose has a specific rotation of + 18.7°. On standing in aqueous solution either of these isomers slowly change rotation (mutarotate) to a specific final equilibrium rotation of + 52.7°. What is the percent of the α and of the β components in the mutarotated equilibrium mixture? (*Hint:* Let x = fraction of α-glucose and $(1 - x)$ = fraction of β-glucose at any time.)

Serum Glucose Analysis: Enzymatic/Colorimetric

OBJECTIVES

1. To gain experience in the use of a photoelectric spectrophotometer.
2. To understand Beer's law as the cornerstone of colorimetric analysis.
3. To gain experience in the application of enzyme catalysis.
4. To understand one procedure for the analysis of glucose in the body fluids.

PRELABORATORY QUESTIONS

1. What is the chemical nature of an enzyme, and what precautions must be observed in its laboratory applications?
2. In what important way is an enzymatic/colorimetric analysis for glucose in blood superior to any of the ordinary chemical oxidation/colorimetric methods in use?
3. What is Beer's law?
4. What is meant by "Beer's law conformance"?

DISCUSSION

The glucose level in the body fluids stays within a predictable range for normal healthy individuals: 50–90 mg/dL. Under certain pathological situations these levels stray from the norms. Glucose analyses are therefore important to the detection of pathological conditions as well as to the monitoring of medical treatments. A number of excellent procedures can be used for glucose analyses. Most of these depend on the ease with which glucose is oxidized. Were it not for the presence of a variety of other reducing agents in body fluids, chemical analyses for glucose would be a snap.

In this experiment you will apply a procedure that nicely circumvents this problem of competing reducing agents. You will use a specific enzyme system that catalyzes the oxidation of glucose only. It is a modification of the "Keston procedure," which utilizes a coupled enzyme system, peroxidase–glucose oxidase (PGO). As indicated in the equations below, this enzyme system oxidizes a mole of water and a mole of D-glucose to 1 mole of D-gluconic acid and 1 mole of hydrogen peroxide. Simultaneously, the hydrogen peroxide produced quantitatively oxidizes an internal indicator, *o*-dianisidine, to a brown material (probably dehydro-*o*-dianisidine). The absorbance of this brown product is directly proportional to the original concentration of glucose and is measured spectrophotometrically at 425–475 nm.

CHO–CHOH–HOCH–CHOH–CHOH–CH_2OH (D-Glucose) + O_2 + H_2O $\xrightarrow[37°/30\ min]{PGO}$ CO_2H–CHOH–HOCH–CHOH–CHOH–CH_2OH (D-Gluconic acid) + H_2O_2

D-Glucose

D-Gluconic acid

H_2O_2 + *o*-Dianisidine (colorless) ⟶ Dehydro-*o*-dianisidine (brown (425–475 nm)) + $2H_2O$

H_2N, NH_2, CH_3O, OCH_3; HN, NH, CH_3O, OCH_3

o-Dianisidine
(colorless)

Dehydro-*o*-dianisidine
(brown (425–475 nm))

Results can then be quantitated by the use of the Beer–Lambert law:

$$\text{absorbance} = -\log\left(\frac{I_t}{I_0}\right) = Alc,$$ where:

- I_t is the intensity of transmitted light (reads on meter as percent transmittance)
- I_0 is the intensity of incident light (set at 100% at start of experiment)
- l is the length of the light path (cm) (1 cm for Spectronic 20 tubes)
- c is the concentration of the colored solute (any units, as long as they are used consistently)
- A is a proportionality constant for a specific colored solute in a specific solvent at a specific wavelength

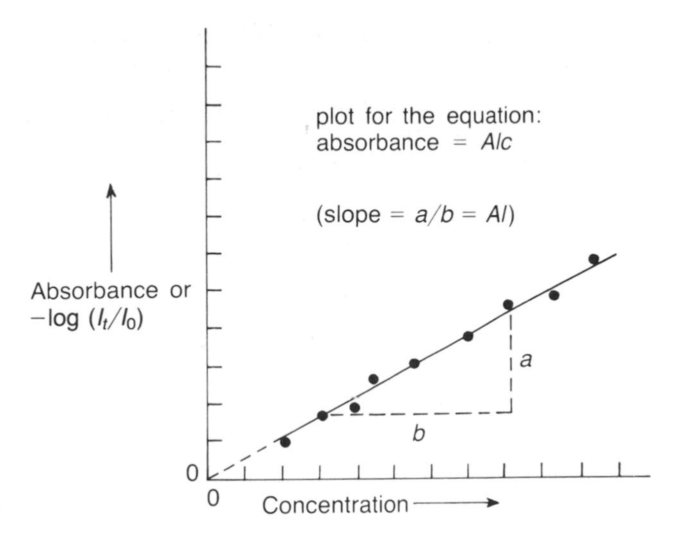

Figure 27.1

The Spectronic 20 spectrophotometer gives direct dial readings of both absorbance and percent transmittance ($I_t/I_0 \times 100$). Moreover, l, the length of the light path in the absorbing solution, is a constant very close to 1 cm. The colorimeter tubes are manufactured to be as uniform in light path as possible.

It should follow, then, that the only variables in the Beer–Lambert equation are absorbance and concentration, c. By plotting absorbance vs. concentration, as in Figure 27.1, one gets a straight line with a positive slope of Al.

Whenever the plot of absorbance vs. concentration gives a straight line, the system is said to "conform to Beer's law," and colorimetry is indicated as an acceptable analytical application for the system. If, on the other hand, a straight line does *not* result when such a plot is drawn, there is no Beer's law conformity, and colorimetry is of questionable value as an analytical application for that particular system. Colorimetry and the application of the Beer–Lambert law are widely used in analyses of body fluids.

Use of the Spectronic 20 Spectrophotometer

(See Figure 27.2.)

1. Turn on the power (left front knob) and allow the instrument to warm for 15 minutes. Set for the desired wavelength (425 nm) with the top right knob. When thermal equilibrium has been attained, proceed as follows.
2. Set to 0 transmittance without a tube in the cuvette holder by adjusting the front left knob. Be sure that the holder lid is closed!
3. Set the spectrophotometer for 100% transmittance by inserting the reagent blank—etched line forward—in the cuvette holder (top left) and turning the right front knob. The colorimeter tube (cuvette) should be about three-fourths full and wiped clean and dry on the outside. Save the reagent blank since the 100% transmittance setting should be reestablished each time you return to use the instrument. Instrument settings drift with changing temperature. Always be sure that the lid over the cuvette is securely closed before taking a reading because external light also interferes with readings.

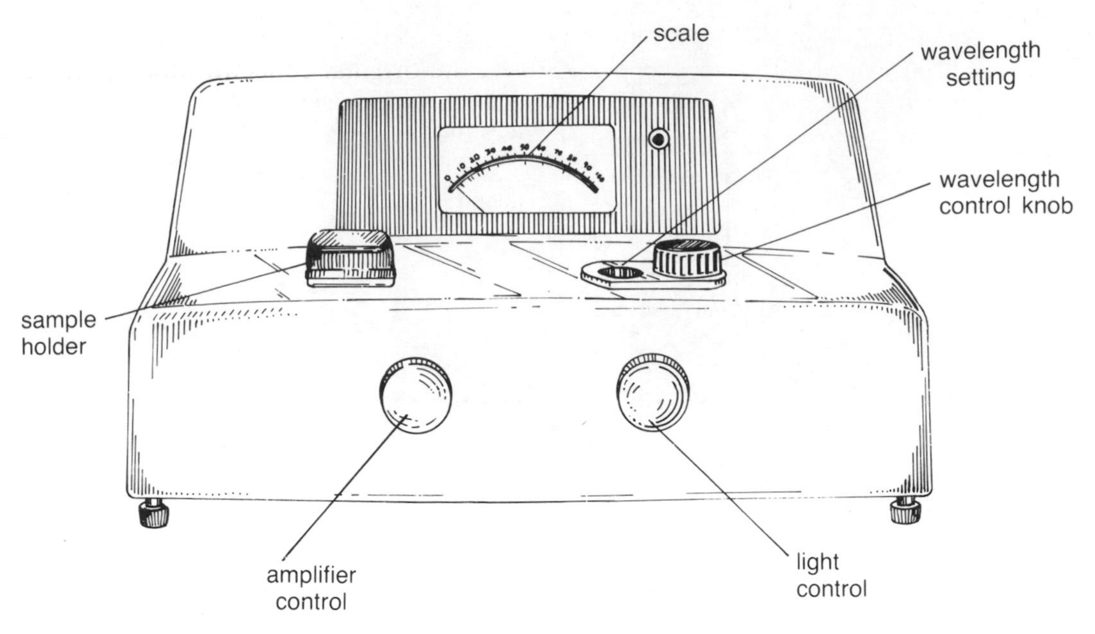

Figure 27.2

4. When taking a reading of your sample, simply replace the reagent blank tube with an externally dried and matched tube that is about three-fourths filled with your sample. Secure the cover and read the absorbance and the percent transmittance directly from the dial.
5. Remove the colorimeter tube, close the holder cover, and leave the spectrophotometer on, with settings intact, since others will be using it for measurements on the same chemical system.

MATERIALS NEEDED

Select a lab partner. You will work as a team, but each of you will analyze an unknown.

Obtain a vial of peroxidase—glucose oxidase enzyme (PGO), one vial per team. Each of you should check out an unknown serum glucose specimen (simulated) and two spectrophotometer cuvettes. Also pick up a 1 mL pipet graduated in $\frac{1}{100}$ mL, a disposable extraction column (SEP-PAK), and a 10 mL syringe.

EXPERIMENTAL PROCEDURE

Your unknown specimen will have a glucose concentration that falls within the limits usually found in medical experience. It may be hypoglycemic, normal, or hyperglycemic. This procedure calls for a twentyfold dilution of the specimen (part F) because of the sensitivity of the analysis. For this reason, the standard solutions provided in the laboratory have been prediluted twentyfold to expedite the analysis. In other words, the standard labeled 100 mg/dL is really only 5 mg/dL. However, since specimen and standards are diluted proportionately, the original concentrations (before dilution) are representative and are used.

To minimize confusion, this procedure is presented in seven parts. The first five direct the construction of a calibration curve from standard glucose solutions. The last two direct the analysis of your unknown.

A. Preparation of the Combined Enzyme–Color Reagent

For this preparation, use an amber bottle of 150 to 250 mL capacity if one is available. The colored glass protects the light-sensitive reagent and prolongs its life. If amber or brown containers are not available, use a 125 mL Erlenmeyer and keep it covered with a towel when not making transfers. Of course, it is important that the glassware be clean!

Add the contents of one PGO capsule to 100 mL of distilled water in the container selected above and swirl gently until solution is complete. To this, add 1.6 mL of the *o*-dianisidine dihydrochloride solution. Mix, cover, and shield from direct light. This reagent is enough to tide your team through these analyses.

B. Preparation of Glucose Standards

Number seven clean 10 mL test tubes and set up a water bath (a 400 mL beaker half filled with water and heated to 37°C). One set of standards is required of each laboratory team. Prepare the glucose standards from the standard solutions provided in the laboratory and according to the following chart. Mix the standard, water, and enzyme–color reagent directly in the numbered test tubes. (Note that this is a tenfold dilution of the already twentyfold diluted reagent standard. This is done to make quantitative detection possible at low ppm.)

Tube Number	Amount & Identity of Standard	mL Water	mL Enzyme–Color Reagent	[Glucose] (mg/dL)
1 (blank)	________________	0.5	4.5	0
2	0.50 mL of 50 mg/dL	0.0	4.5	50
3	0.50 mL of 60 mg/dL	0.0	4.5	60
4	0.50 mL of 70 mg/dL	0.0	4.5	70
5	0.50 mL of 80 mg/dL	0.0	4.5	80
6	0.50 mL of 90 mg/dL	0.0	4.5	90
7	0.50 mL of 100 mg/dL	0.0	4.5	100

C. Incubation of the Standards

Mix the six standards just prepared by swirling gently, or better yet, by grasping the tube firmly at the rim with the fingertips of one hand, and while thus suspending it vertically, gently slapping the bottom of the test tube with the forefinger of the other hand. When the stirred water bath is well adjusted to 37° +/− 1°, place all six standards in the bath. *Maintain this incubation temperature for 30 minutes.*

D. Determination of the Absorbance of the Standards

If the spectrophotometer has not already been temperature equilibrated, turn it on (left front knob on the Spectronic 20) and let it warm up for 15 minutes. Set the wavelength at 425 nm (top right dial), and without a cuvette in the holder and with the holder cover closed, set for 0 transmittance with the left front knob. Transfer your blank (tube 1) to a cuvette, wipe the outside of the cuvette dry and clean, and insert it— etched marker out—in the holder. Close the lid and set the transmittance at 100% by use of the right front knob. Remove and *save* this blank for use each time you return to the spectrophotometer for absorbance readings. Close the holder lid. The dial should read 0% transmittance again. If it doesn't, reset to 0 with the left front knob, insert the blank again, and with the right front knob, reset again to 100% transmittance. The instrument should now be ready for use. This instrument adjustment should be repeated each time you return to use the spectrophotometer since it will drift with changes in temperature.

Transfer the contents of tube 2 to a second cuvette, wipe as before, and insert, etched line forward in the holder. Close the lid and read, and record on your data sheet, the absorbance and the percent transmittance directly from the dial. Empty the cuvette (or transfer back to tube 2), rinse it, shake dry, and pour the tube 3 standard into it. Wipe clean and take its absorbance and percent transmittance. Repeat this procedure for standards 4, 5, 6, and 7.

E. Construction of Calibration Curve

Using the grid provided on the data sheet, plot the absorbance readings vs. the concentration (last column in the chart in part B), as illustrated in Figure 27.1. Draw the best *straight* line through these points. If your points won't reasonably accommodate a straight line, repeat parts B through E with greater attention to detail. This system *should* obey Beer's law, and if a clearly defined slope is not indicated, your subsequent analyses can have no validity. Calculate the slope of your line. (See the discussion section for instructions.)

F. Decolorization and Dilution of the Specimen

Prime your disposable extraction column by charging it with 5 mL of dry methanol, slowly forcing the solvent through the column. Follow this with a 10 mL water wash (as described in Experiment 21). Transfer your unknown serum–glucose specimen to the syringe, and slowly elute at a rate of two drops per second into a small, clean, dry test tube. You have now removed the similated bile pigments that might interfere with the subsequent spectrophotometric analysis.

With a clean, rinsed, 1 mL pipet, transfer exactly 1.0 mL of your decolorized specimen to a clean 25 mL Erlenmeyer flask (or similar small receptacle). Add exactly 19.0 mL of distilled water and mix. This is the twentyfold dilution of the specimen alluded to earlier.

Flush your extraction column with 10 mL of water and then with 5 mL of methanol into a sink. The column may be used again with Experiments 36 and 37.

(*If whole blood or plasma is to be analyzed for glucose by this procedure,* *the formed elements and proteins must first be removed by hemolysis, flocculation with $Ba(OH)_2/ZnSO_4$, and filtration or centrifugation.*)

G. Determination of the Absorbance of Your Unknown

Prepare to analyze three aliquots of your decolorized unknown by charging each of three clean 10 mL test tubes with:

1. 0.50 mL of decolorized diluted unknown solution (from part F).
2. 4.5 mL of your enzyme–color reagent (from part A).

Be sure that your pipet is rinsed with water and then with the solution to be transferred before measuring out these quantities!

Mix and incubate at 37° +/− 1° for 30 minutes. After incubation, transfer the contents of one tube to a clean, rinsed cuvette. Wipe the cuvette clean.

Set the spectrophotometer for 0% and 100% transmittance with your blank, as was done for the standards in part D. Check to make sure that the instrument is still set at 425 nm. Some people unconsciously twirl dials.

Determine the absorbance and percent transmittance of your sample and record.

Repeat these steps with your other two aliquots. Average your results.

From the slope of your calibration curve and Beer's law, calculate the glucose concentration of your specimen in terms of milligrams of glucose per deciliter (mg glucose/100 mL of serum). If your unknown represents severe hyperglycemia, you will have to extrapolate your standard curve to establish glucose concentration. Rinse and towel dry the cuvettes, the syringe, and the pipet and amber bottles (if used) and return to the storeroom. Complete the data sheet.

DATA SHEET

Name: ______________________________ Lab Section: ______ Date: __________

SERUM GLUCOSE ANALYSIS: ENZYMATIC/COLORIMETRIC

Unknown # __________ has a glucose concentration of __________________

Absorbance of standard solution: __________ __________ __________ __________ __________ __________

(50 mg/dL) (60 mg/dL) (70 mg/dL) (80 mg/dL) (90 mg/dL) (100 mg/dL)

Abs (unknown) __________; __________; __________ Average absorbance __________ (Plot this point on the graph and draw a circle around the point)

% transmittance (unknown) __________________ (average) Slope __________

Absorbance: 1.4, 1.3, 1.2, 1.1, 1.0, 0.9, 0.8, 0.7, 0.6, 0.5, 0.4, 0.3, 0.2, 0.1, 0.0

0, 20, 40, 60, 80, 100, 120, 140, 160, 180, 200

Concentration (mg/dL)

QUESTIONS

1. Why was it important to decolorize the serum specimen before subjecting it to the glucose analysis?

2. Recall that you set the spectrophotometer for absorbance at 425 nm, the wavelength where the brown-colored, oxidized *o*-dianisidine exhibits maximum absorbance. Suppose that you were planning a spectrophotometric analysis of another colored system for which maximum absorbance was not available to you. How might you determine experimentally on your spectrophotometer an appropriate wavelength setting to use? (Remember that the top right dial knob can be used to scan the entire light absorption range within the instrument's capability.)

3. What is the normal serum glucose range expressed in mg/dL?

4. a. What is the difference between whole blood and blood serum?

   b. What is the difference between blood plasma and blood serum?

EXPERIMENT 28

Chromatography of Lipids: Lecithins and Cholesterol

OBJECTIVES

1. To learn to isolate a lipid from a natural source.
2. To establish chromatographic properties of some lipids.
3. To be able to identify an unknown lipid on the basis of its chromatographic properties.

PRELABORATORY QUESTIONS

1. What is meant by R_f value?
2. Of the lipids illustrated in the discussion, which is the most polar?
3. Of the three solvents, H_2O, $(CH_3)_2CHOH$, and CCl_4, which is the most lipophilic (hydrophobic)?
4. For your developing chamber you are asked to construct a wick from filter paper. What is the purpose of this wick?
5. Why is it important to keep the spot diameter small when doing TLC?

DISCUSSION

Thin layer chromatography (TLC) is a rapid technique for the qualitative separation and identification of mixtures. With TLC the progress of an organic reaction can be nicely monitored by periodically spotting a TLC plate with a drop of the reaction mixture and developing to observe the appearance of spots for products and the disappearance of reactant spots. One can easily run a TLC on a final reaction product to determine whether impurities are present.

In this experiment we will determine the chromatographic properties (R_f values) for two common lipids, lecithin and cholesterol, for a specific solvent system. Then we will use these properties to analyze an unknown extract for the presence or absence of these two lipids. We will also extract the carotenoids from tomato catsup and establish their chromatographic properties.

Some of the lipids that you encounter in this experiment are:

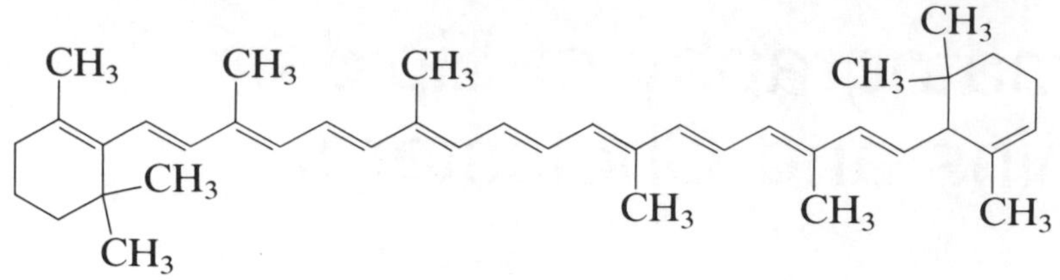

α-Carotene

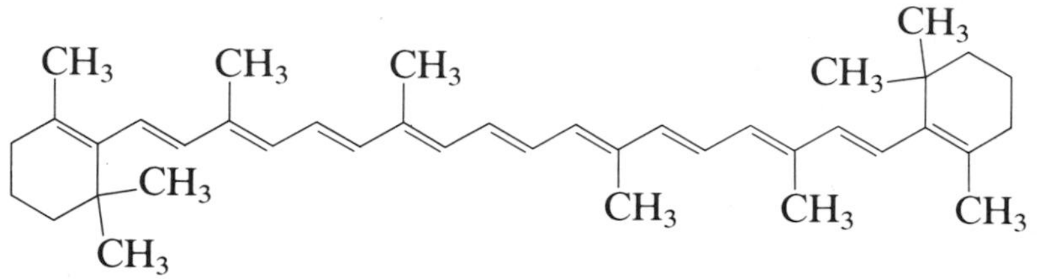

β-Carotene

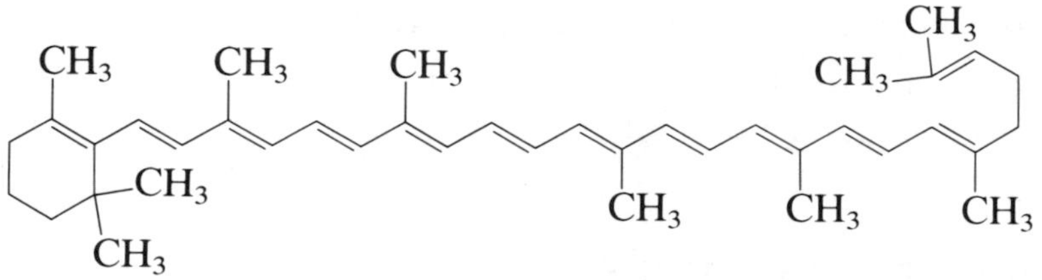

γ-Carotene

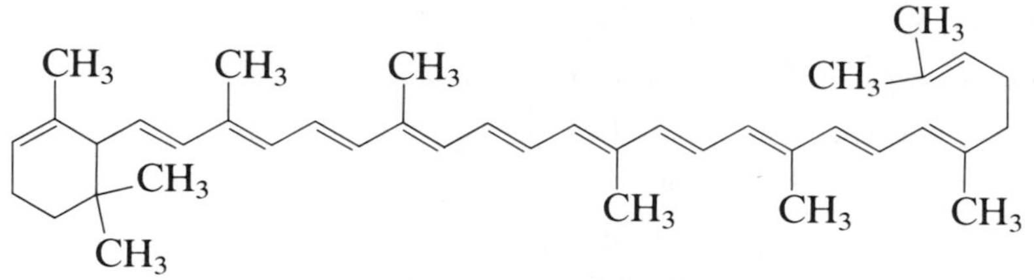

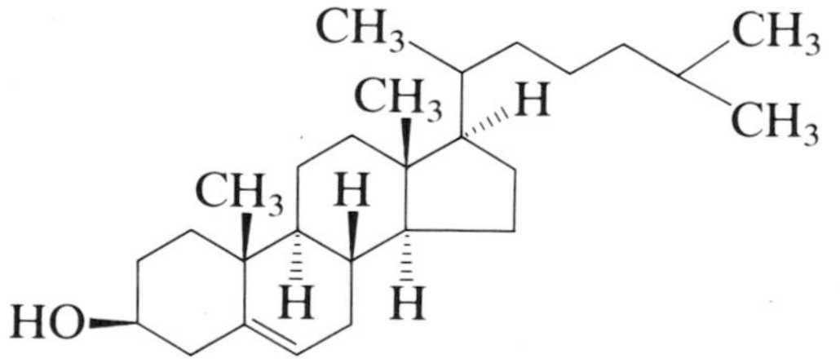

δ-Carotene

Cholesterol

$$
\begin{array}{l}
\mathrm{CH_2O{-}\overset{\overset{\displaystyle O}{\|}}{C}{-}(CH_2)_{16}{-}CH_3} \\
\mathrm{HCO{-}\overset{\overset{\displaystyle O}{\|}}{C}{-}(CH_2)_7{-}CH{=}CH{-}CH_2{-}CH{=}CH{-}(CH_2)_4{-}CH_3} \\
\mathrm{CH_2O{-}\underset{\underset{\displaystyle O^-}{|}}{\overset{\overset{\displaystyle O}{\|}}{P}}{-}O{-}CH_2{-}CH_2{-}\overset{+}{N}(CH_3)_3}
\end{array}
$$

a Lecithin
(a phosphatidylcholine)

Vitamin A (retinal)

By definition, lipids are water-insoluble compounds of biological origin. Methylene chloride, CH_2Cl_2, is an excellent general-purpose solvent for lipids and is our solvent of choice for the extraction of the carotenoids from the biological matrix that is catsup.

Lipids include the four fat-soluble vitamins: A, D, E, and K. The triacyl glycerides, the fatty acids, the phosphoglycerides, the sex hormones, and cholesterol are just a few other common biological examples of the lipids. A close inspection of the structures shown above reveals a definite structural relationship between the carotenoids and vitamin A.

Thin layer chromatography (TLC) is a differential partition process. Solutes in solution are placed (spotted) on the starting line of a chromatography strip (Figure 28.1). This strip is coated with a stationary solid adsorbant, a material through which a liquid will flow by capillarity, but which nonetheless exhibits an attraction toward solute materials dissolved in the slowly flowing solution. In this experiment we will use commercially available silica gel strips. The silica gel is essentially a precipitated form of silicic acid, H_2SiO_3, and is quite polar in nature. As a consequence, solutes being transported over its surface will be attracted to the surface (selectively adsorbed) in direct proportion to their own intrinsic polarity. The greater the polarity of a passing solute, the more hindered will be its passage over the polar stationary TLC strip. The greater the solute polarity, the more the rising

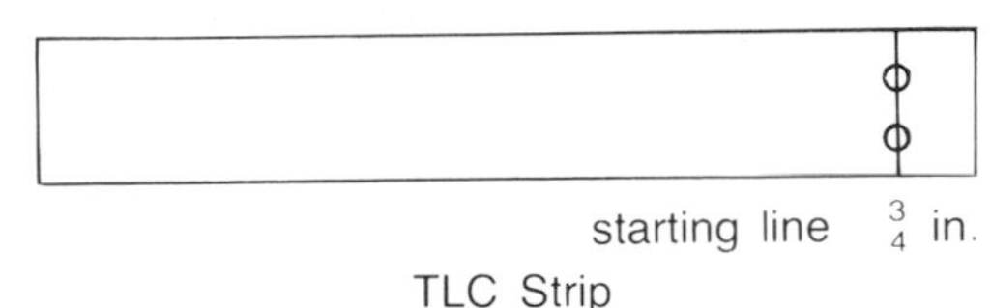

Figure 28.1 TLC Strip

solvent front will leave it behind. Thus the R_f value for that solute will be relatively low. By definition:

$$R_f = \frac{\text{distance traveled by the solute (spot)}}{\text{distance traveled by the solvent (solvent front)}}$$

Solvents used for developing can be tailor-made for optimum separations of different mixtures. For our particular purposes, the best separations have been observed for the three-part mixture—14 parts of chloroform, 6 parts of methanol, and 1 part of water—so that is the developing solvent provided. You are encouraged to try other solvents or mixtures in an effort to improve results beyond what this mixture provides.

For our developing chamber we will use a 400 mL beaker covered with an aluminum foil cap. The developing solvent is introduced first, to a depth not to exceed 0.5 cm. A curved piece of filter paper is adjusted to stand upright along one wall as illustrated in Figure 28.2. This serves as a wick to accelerate the equilibration of liquid and vapor phase pressures within the chamber. Such equilibration is necessary to minimize solvent evaporation from the chromatography strip. Since the more volatile solvents will evaporate the fastest (in the absence of prior equilibration), it should be apparent that under such conditions the character and adjusted polarity of the developing solvent would change as it rose on the strip. This, of course, would alter solute migration with time and make R_f determinations unreliable.

Visualization is the detection and identification of solute spots. Highly colored solutes can be seen as spots without chemical or physical accentuation. Highly conjugated molecules will absorb ultraviolet light and can often be detected by illuminating the TLC strip with an ultraviolet lamp. Selected acid-base or redox indicators can be sprayed on a strip to colorize other spots. Perhaps the most generally used form of visualization is to expose the developed strip in an iodine chamber. Iodine vapors are readily absorbed by organic compounds and can be seen as brown to violet spots wherever a solute spot has developed. In iodine visualization, however, one should always quickly draw a light circle around each spot with a pencil since once removed from the iodine jar, the spots slowly disappear as the absorbed iodine sublimes.

For the purposes of this experiment, however, spots will be visualized as a blue phosphomolybdate complex. The developed TLC strip is sprayed lightly and evenly with a solution of phosphomolybdic acid ($20\ MoO_3 \cdot 2\ H_3PO_4 \cdot$

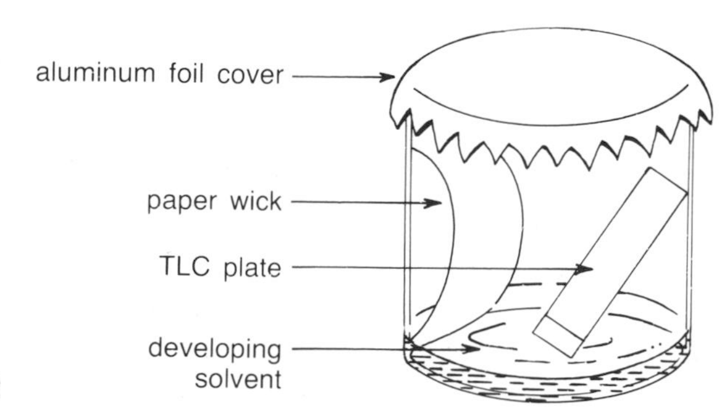

Figure 28.2

48 H_2O; *MW* 3940), dried, baked in an oven briefly, and finally exposed to ammonia vapors. When baked, the lipids in the spots complex with the phosphomolybdate and can be seen as blue to violet spots on a yellow-green background. When the strip is then exposed to NH_3 vapors, the yellow-green background decolorizes and the blue spots are more clearly distinguishable.

MATERIALS NEEDED

From the storeroom obtain a small vial containing a methylene chloride solution of a mixture of lecithins, a mixture of cholesterol, or both lecithins and cholesterol. All other chemicals and materials needed will be available in the laboratory unless your instructor notes otherwise.

EXPERIMENTAL PROCEDURE

It is suggested that you proceed with each part of this experiment in the order given; this will maximize results and save time.

A. Preparation of the Developing Chamber

Fit a piece of filter paper along the wall of a clean, dry 400 mL beaker as shown in Figure 28.2. Add about 25 mL of the developing solvent, $CHCl_3/CH_3OH/H_2O$ (14-6-1), to the beaker. The solvent should be about 0.5 cm deep in the beaker. Cover this chamber by tightly securing a 6 × 6 inch piece of aluminum foil over its top.

Allow this developing chamber to stand quietly and attain a liquid–vapor equilibrium while you proceed to extract the carotenoids from catsup.

B. Extraction of Carotenoids from Catsup

In a 100 mL beaker mix about 5 mL of tomato catsup with 10 mL of methanol and stir for a minute or so to extract as much water as possible from the biochemical matrix. Separate the methanol extract by filtration through a fluted filter (Figure 28.3), and then dry the residue by pressing the filter paper and its contents several times between dry paper toweling. When as dry as possible, place the cake of catsup residue in a small dry beaker or flask, and stir it with 5 mL of methylene chloride. You can see the colored carotenoids appear as you stir for about 5 minutes. Decant or filter the colored extract into a dry, covered container and save for the TLC analysis in parts C–E.

C. Spotting the Sample on the Chromatogram

If capillary applicators are not provided, make four of them by drawing them from blood sample capillaries. To do this, rotate a blood capillary in a soft blue flame until very soft. Then quickly remove from the flame and draw the softened glass about 4 inches (Figure 28.4). Broken in half, this provides two good capillary applicators. You will want a separate and uncontaminated

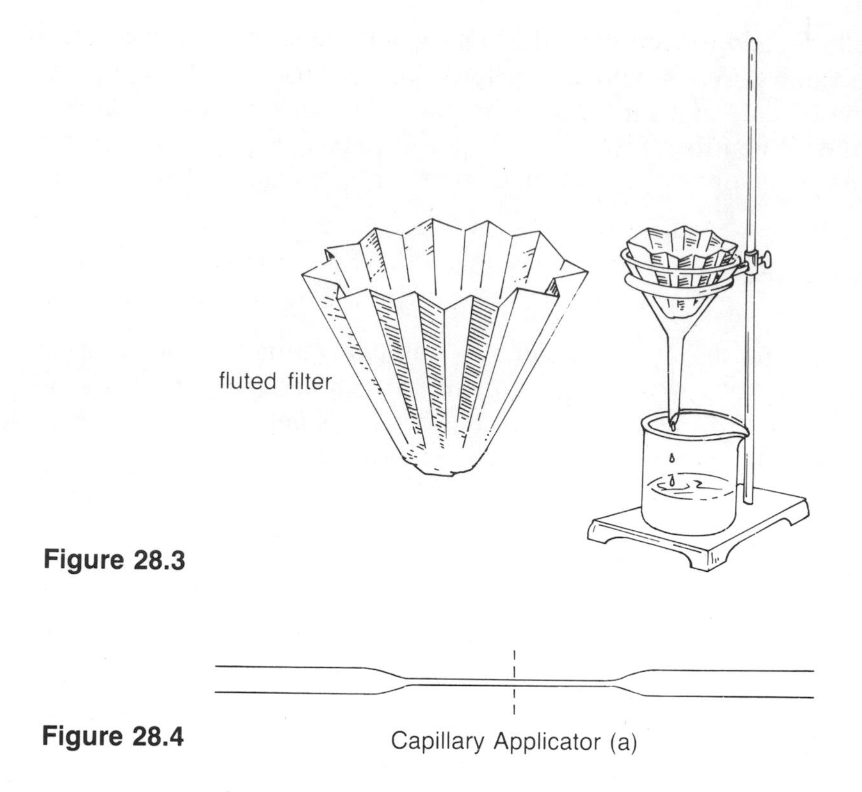

Figure 28.3

Figure 28.4

capillary applicator for each of the four solutions investigated in this experiment.

Get a TLC strip—handling it only by what you should designate as its top end. Fingerprints develop as spots, too. Be especially careful not to contaminate the rough silica gel surface in any way. Place the strip, silica gel side up, on a piece of clean paper toweling and *lightly* mark the starting line with a pencil about 2 cm from the designated bottom end of the strip. See Figure 28.1. Do not press the pencil heavily since, if the gel surface is scratched off at this starting line, the rising solvent will be unable to cross the starting line. Do *not* use a ballpoint pen!

Touch the fine tip of a capillary applicator to the extract of carotenoids prepared in part B. Now lightly touch the applicator tip to the starting line *about one-quarter of the way in from the left edge.* Remove it quickly—as soon as a spot about the size of a pinhead has appeared. If the spot is allowed to grow to a diameter of greater than 2 to 3 mm, accurate R_f values are harder to establish. In the same way, touch a second spot at the *center* of the starting line, and a third *one-quarter of the way in from the right edge* of the strip. Allow these spots a minute to dry, and then make a second application at the exact center of your number 2 and 3 spots on the starting line. Again, do not allow these applications to enlarge the original spot. Finally, make a third application on the right-hand number 3 spot. By preparing these applications in three different concentrations, you will then be in a position to establish the optimum spot concentration for future determinations for this particular extract. If, for example, a spot concentration is too weak, minor components of the mixture might be imperceptible in the visualization. On the other hand, too high a solute spot concentration may exceed the solvent's capacity to carry it all when it first passes through the spot, and streaking will be observed in the visualization.

D. Developing the Chromatogram

Take special care to work rapidly with a minimum exposure of the open developing chamber to the air and a minimum of movement of the chamber.

Remove the cover and carefully place the TLC strip in the chamber such that the bottom of the strip enters the solvent pool vertically and without splashing, and such that when the strip is then allowed to lean against the chamber wall, you will be able to observe the rise of the solvent front on the gel surface. It is very easy to immerse the strip in the solvent pool such that an uneven front arrives at the starting line. When this happens, of course, comparative R_f values are of minimal reliability. Quickly close the chamber with the foil—as securely as possible and without disturbing the surface tranquility of the solvent pool. At this juncture you would be well advised to get another student to help by holding the beaker securely to the bench top while you adjust and seal the cover.

When the solvent front has risen about 10 cm, or when it appears that its upward progress is almost arrested (about 15 minutes for this system), remove the strip from the developing chamber and quickly, before it disappears, mark the solvent front lightly with a pencil line. Allow the chromatogram to dry, face up, on a piece of dry paper toweling for 5 minutes. Don't forget to put the cover back on the developing chamber so that equilibration can be attained before you use it next.

E. Visualizing the Chromatogram

Be especially careful in spraying a chromatogram. An atomizer that produces a very fine mist is necessary. The developed TLC strip must be uniformly moistened by the spray, and yet *not* moistened to the extent that the applied liquid will run on the plate. When too much spray is applied, spots will sometimes merge or even run to the edge of the strip, and of course the visualization then becomes meaningless.

Suspend your chromatogram by a clip (in a hood if available). This is a good place to summon an instructor for guidance. Spray the chromatogram with the phosphomolybdate solution until uniformly dampened but not soaked. Allow the chromatogram to dry for a couple of minutes by placing it, gel side up, on a piece of paper toweling.

Place your chromatogram on a large watch glass, or on the bottom of an upturned beaker, and put it in the oven (maintained at 100 to 110°) for $2\frac{1}{2}$ to 3 minutes. This should be sufficient for blue spots to form on a yellow-green background. If the oven has cooled due to frequent opening of the door, more time is required for the visualization. However, timing is critical since ultimately heat will convert spots and background to a uniform dark blue and spots will not be detectable. Ideally, the oven should have a window through which you might monitor the visualization for optimum baking time.

Remove the chromatogram from the oven and hold it over an opened bottle of concentrated ammonia in the fume hood (if available) to bleach away the yellow-green background. The blue spots should now be clearly discernible. Lightly pencil a circle around each spot, mark the center of density of the spot, and measure its migration (in millimeters) from the starting line.

Calculate the R_f value for each spot, and on your data sheet diagram your chromatogram as Plate I, labeling each spot with its correct R_f value.

F. Analysis of an Unknown Lipid Solution

On a fresh TLC strip draw a starting line as described in part C and spot it with the known cholesterol reference standard on the left, with your unknown lipid solution in the center, and with the known lecithin solution on the right Two applications should be about right, but you are encouraged to run simultaneously a trial strip with your unknown at three different application concentrations in order to establish the optimum conditions for you and your particular unknown. Please remember that chromatography is an art as well as a science, and each investigator must find his or her own best operating conditions.

As you did in part D, develop your chromatogram in the developing chamber, taking care to avoid disturbing the liquid–vapor equilibrium in the chamber. When the solvent front has risen about 10 cm, remove the strip, quickly mark the solvent front, dry in air for 5 minutes, spray evenly and lightly with phosphomolybdate solution as in part E, dry, bake in the oven, decolorize the background with NH_3 vapors, and mark and measure migration distances to calculate R_f values. On your data sheet, diagram your *best* chromatogram (you may have to do several of them) as Plate II and label it with calculated R_f values. Establish whether your unknown contains (1) cholesterol, (2) lecithins, or (3) both cholesterol and lecithins. Answer the questions on the question sheet and return the unknown vials to the storeroom.

DATA SHEET

Name: ______________________________ Lab Section: ______ Date: ____________

CHROMATOGRAPHY OF LIPIDS: LECITHINS AND CHOLESTEROL

Carotenoids in Catsup

Diagram the carotenoid separation as Plate I (indicate your calculated R_f values for each spot).

Plate I

Identification of Lecithins and/or Cholesterol in Unknown
____

Diagram as Plate II your best results in chromatographing the lecithins, cholesterol, and unknown. Label all observed spots with their calculated R_f values.

Plate II

Unknown # __________ contained ______________________________

QUESTIONS

1. Why is it important not to add too much sample to a spot on a TLC plate?

2. Suppose that you had used the developing solvent 25% CH_2Cl_2/75% n-C_6H_{14} and you found that the relatively polar lecithins were barely leaving the starting line. Using the same solvent pairing, how might you adjust the developing solvent to get a better lecithin migration?

3. Of the lipids illustrated in the discussion section in this experiment, which would you expect to have the lowest R_f value, assuming that all are quite soluble in the developing solvent used?

4. Recall that iodine number is a measure of the degree of unsaturation in an organic molecule. (a) How many moles of I_2 would be adsorbed by 1 mole of α-carotene? (b) By 1 mole of vitamin A? (See structures in the discussion section.)

5. Write the equation for the complete digestion (hydrolysis) of lecithin. (See the structure for lecithin in the discussion section.)

Amino Acid Analysis: Paper Chromatography

OBJECTIVES

1. To introduce paper chromatography as an analytical tool.
2. To utilize the amphoteric character of the amino acids in their separation and identification.

PRELABORATORY QUESTIONS

1. Define R_f value.
2. Which of the amino acids studied in this experiment is a "basic amino acid"?
3. What is meant by "isoelectric point"?
4. When dissolved in water, aspartic acid, $HO_2CCH_2CH(NH_2)CO_2H$, has anionic character (isoelectric point equals 2.77) and is attracted to a positive electrode. Would you add acid or base to an aqueous solution of aspartic acid in order to bring it to its isoelectric point? (*Hint*: Study the equilibrium found in the discussion.)
5. Define pH.

DISCUSSION

This experiment is very similar in technique to the thin layer chromatography of Experiment 28. The basic difference is in the selection of a highly polar developing solvent and of a cellulose stationary adsorbant. A little reflection will justify these choices.

Amino acids are not only highly polar, but they are also capable of strong hydrogen bonding. A developing solvent rich in water is certainly a logical candidate for the role of carrier solvent. Then the insoluble cellulose, with its high OH density (three hydroxy groups per monomer glucose unit), makes a formidable obstacle course for the ascending amino acids. The greater the H-bonding capability of an amino acid, the greater the difficulty it has in passing over the cellulose surface along with its carrier solvent. As was the case in thin layer chromatography, the stronger the attractive force between solute and stationary phase, the lower the R_f value for that solute.

You will be issued an unknown containing from one to three of the following amino acids. Reference solutions of each of these six amino acids will be provided to establish the chromatographic properties needed to analyze your unknown mixture.

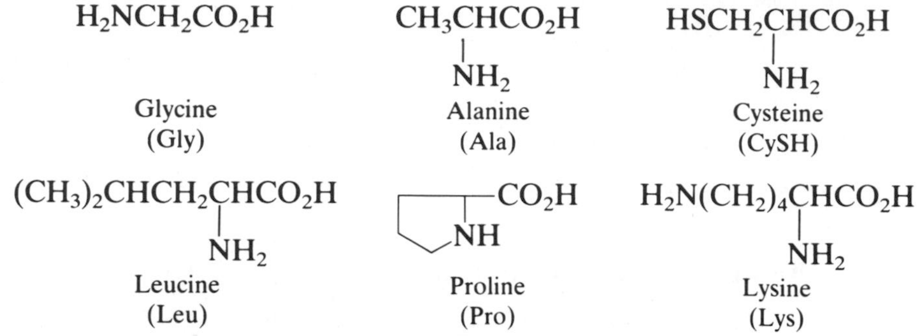

$H_2NCH_2CO_2H$
Glycine
(Gly)

$CH_3CH(NH_2)CO_2H$
Alanine
(Ala)

$HSCH_2CH(NH_2)CO_2H$
Cysteine
(CySH)

$(CH_3)_2CHCH_2CH(NH_2)CO_2H$
Leucine
(Leu)

Proline (pyrrolidine ring: —CO_2H, NH)
Proline
(Pro)

$H_2N(CH_2)_4CH(NH_2)CO_2H$
Lysine
(Lys)

Most of the original work done in paper chromatography involved the use of Whatman No. 1 filter paper (cotton cellulose). Though new chromatographic supports have been developed, filter paper is still widely used, and we will use it here.

Amino acids have a remarkable capacity to alter their polarity when the pH of their environment changes. They are amphoteric molecules; that is, they have both acid and base capacity. Consider the amino acid alanine in water solution:

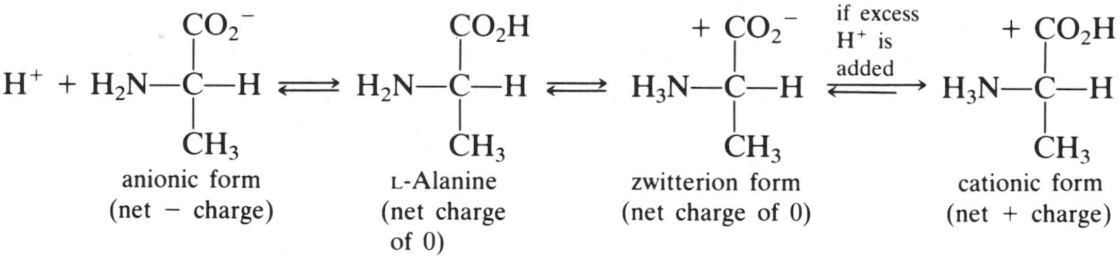

$$H^+ + H_2N\text{—}C(CO_2^-)(CH_3)\text{—}H \rightleftharpoons H_2N\text{—}C(CO_2H)(CH_3)\text{—}H \rightleftharpoons H_3\overset{+}{N}\text{—}C(CO_2^-)(CH_3)\text{—}H \xrightleftharpoons[\text{added}]{\text{if excess } H^+ \text{ is}} H_3\overset{+}{N}\text{—}C(CO_2H)(CH_3)\text{—}H$$

anionic form (net − charge); L-Alanine (net charge of 0); zwitterion form (net charge of 0); cationic form (net + charge)

When alanine is dissolved in water, the equilibrium is shifted to the left such that the anionic form is the predominating solute species. This happens because the carboxylic acid group is slightly stronger as an acid than is the amino group as a base. Another way of saying this is that the —CO_2H group is a stronger acid than is the —NH_3^+ group. At any rate, when an aqueous solution of alanine is subjected to an electrical field, the alanine molecules proceed to migrate spastically toward the anode (positive electrode). Whenever an alanine molecule acquires a negative charge, then, it makes a move toward the anode. Note that the un-ionized form and the zwitterion form would not be affected by such an electric field. Note also that if the pH of the carrier solution is lowered by the addition of acid, a cationic form develops that would naturally be attracted to a cathode. This chameleonlike polarity is the basis for *electrophoretic* separations of amino acids—but that is another story. What is important to this experiment in paper chromatography is that in a solution of any specific pH each amino acid has its own unique polarity, and therefore has its own unique chromatographic properties.

It is worth noting that when the pH of an amino acid solution is adjusted such that the zwitterion form of the amino acid is the predominant species, that pH is called the *isoelectric point*.

Isoelectric Point (pI): That pH at which an ampholyte (an amphoteric substance) has a net charge of zero and is not affected by an electric field, *and* exhibits its minimum water solubility.

Every amino acid has its own specific isoelectric point. Study the zwitterion structure of alanine for a moment. Note that it should be able to form dimers having two salt bridges. These strongly bound dimers are so wrapped up in mutual attraction that they have little interest in water and therefore are not well solvolized by water.

CH_3 — CH — $C(=O)$ — $O^{\ominus}$ ··· $^{\oplus}NH_3$ — $CH(CH_3)$ — $C(=O)$ — $O^{\ominus}$ ··· $H_3N^{\oplus}$ (cyclic dimer)

salt bridge → ← salt bridge

Zwitterionic alanine as a dimer
(all charges are dissipated in salt bridges)

All the other forms of alanine in the equilibrium, on the other hand, attract H_2O more effectively, both by H-bonding and by virtue of their polarity.

Looking at the equilibrium again, it should become apparent that in the case of alanine, the isoelectric point is attained by adding *just enough* acid to the solution to shift the equilibrium from its predominately anionic form to the zwitterion form (center of the equation). If too much H^+ is added, the isoelectric pH is overshot and a cation results.

Ninhydrin (triketohydrindene hydrate) is the best visualization reagent for the detection of α-amino acids. Sprayed lightly and evenly on a dried developed chromatogram and then baked in an oven at 80°C for 5 minutes, it forms colored (usually purple) spots wherever an amino acid is present on the strip. Even the spot colors have some diagnostic significance. Cysteine, for example, gives a brown spot. The chemical reaction between ninhydrin and α-amino acids is well known but quite complex.

MATERIALS NEEDED

Pick up a numbered vial containing an unknown amino acid mixture. All other materials needed will be available in the laboratory.

EXPERIMENTAL PROCEDURE

A. Preparation of the Chromatogram

Being very careful to keep the paper free from fingerprints and dirt, procure (or cut out of filter paper) a 10 × 20 cm strip of Whatman No. 1 filter paper. Draw a light starting line 1.5 cm from one of the long edges, and along this line lightly mark eight starting points equidistant from one another. Along the top edge, in light small penciled letters, label the eight starting points for the six knowns and two unknowns, namely: Pro, Ala, Gly, Leu, Lys, CySH, U_1, and U_2, as in Figure 29.1.

Obtain four blood capillaries and make eight capillary applicators as instructed for Experiment 28—one for each of your eight solutions. Be careful not to mix them up.

Spot each of the reference compounds on the appropriate starting dots, and then spot your unknown at U_1 and U_2. Make the spot diameters no larger than 3 mm and apply three applications per spot. Then, on your second unknown spot, U_2, make two additional applications. This first run will tell you whether the spot concentration is too great (streaking) or too little (dim). (You adjust your spot applications accordingly on your second run.)

Allow the spots a few minutes to dry, and then, taking care to minimize contamination, staple the short ends of the chromatogram together such that its paper edges *do not* touch each other. See Figure 29.2.

B. Developing the Chromatogram

In a clean, dry 800 (or 1000) mL beaker add developing solution (butanol:acetic acid:water; 4:1:5) to a depth of 0.5 cm. Cover with a watch

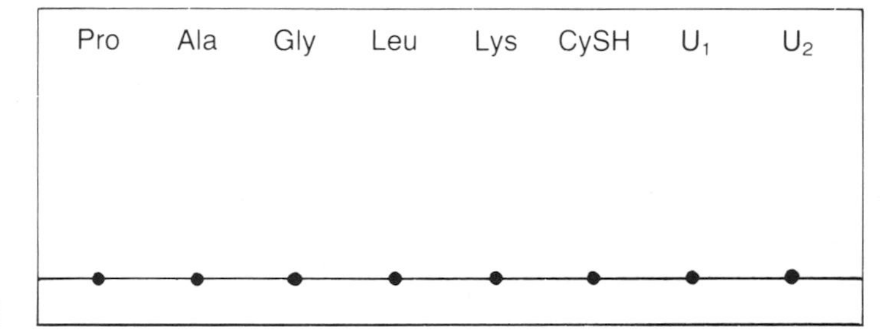

Figure 29.1

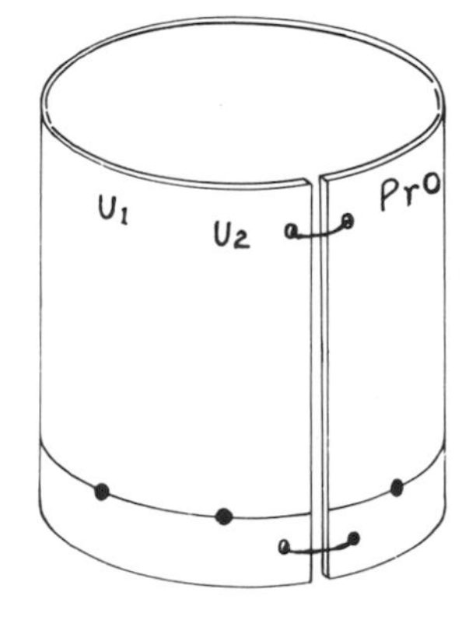

Figure 29.2

glass or an aluminum foil cover (see Figure 28.2) and allow 5 minutes for liquid–vapor equilibration. Insert the cylindrical chromatogram carefully and vertically into the developing chamber (beaker) such that it stands in the center of the beaker and nowhere touches the beaker walls. Holding the beaker firmly on the bench top, quickly and tightly close the chamber again. Keep closed until the completion of the developing and avoid agitation that might affect the tranquility of the solvent pool.

Allow the chromatogram to develop until the solvent front is about 2 cm from the top of the cylinder. Remove the chromatogram cylinder from the chamber and quickly—before it disappears—mark the solvent front. Remove the staples and allow the chromatogram to dry for a few minutes on a clean, dry glass surface (bottom of an upturned beaker).

C. Visualization of the Chromatogram

Initial your chromatogram and hang it by a couple of paper clips on the wire strung for that purpose in the hood. Spray lightly and evenly with ninhydrin. Allow a few minutes to dry, and then bake for 5 minutes on a clean surface in an oven preheated to 80°C. Remove from the oven, and lightly and exactly pencil in the outline of each visible spot. Note any differences in spot color.

Measure the distance of the centers of color density for each spot from the starting line, and also measure the distance the solvent front traveled. Calculate the R_f values for each spot (see Experiment 28).

Staple your *best* (you may need to run several trials to be sure of your results) chromatogram to the data sheet and label all spots with their experimentally determined R_f values. Identify the composition of your unknown amino acid(s) solution.

DATA SHEET

Name: ______________________________ Lab Section: ______ Date: __________

AMINO ACID ANALYSIS: PAPER CHROMATOGRAPHY

Attach your best chromatogram here, completely labeled with R_f values:

Unknown # ________ is a solution of ______________________________

QUESTIONS

1. Of the six amino acids listed as possible components in your unknown, which would have the most *cationic* character when dissolved in water, and why?

2. a. Which of the six amino acids is without chirality?

b. Which should be the least water soluble?

3. Write the structure for proline as a zwitterion at its isoelectric point.

4. Which of the six amino acids studied today is/are "essential amino acids"?

5. Which of our six amino acids should have the best hydrogen bonding capability at its own isoelectric point?

6. How is a chromatogram affected when the liquid–vapor equilibrium of the developing solvent in the chamber is not attained?

7. In Experiment 28, silica gel is the designated stationary adsorbant, whereas in this experiment it is cotton cellulose. In what way do these two stationary phases differ in their manner of interfering with the flow of solutes over their surfaces?

The Physical and Chemical Properties of Proteins

OBJECTIVES

1. To gain familiarity with some fundamental properties of protein.
2. To understand the difference between 1° and 2° protein structures.
3. To understand denaturation and its common causes.
4. To understand and use several qualitative tests for proteins.

PRELABORATORY QUESTIONS

1. Draw the structure of the simple dipeptide glycylalanine and underline the peptide (amide) bond.
2. What happens to a protein when it is denatured?
3. What is a "salt bridge"? Illustrate.
4. Write an equation for the digestion of the dipeptide glycylalanine as catalyzed by a dipeptidase in the small intestine where the pH is about 8.

DISCUSSION

Proteins are extremely complex and delicate molecules and are difficult to work with because they are so easily denatured. This is a nonquantitative experiment in descriptive chemistry designed to demonstrate some of the physical and chemical stresses protein *cannot* withstand. The experiment will provide a substantial insight into the nature of proteins as well as a battery of tests useful for subsequent studies in physiology and biochemistry.

In this experiment you will observe a wide variety of protein reactions. These same sensitivities are characteristic of the body tissues and enzymes when they encounter similar changes in their chemical environment. An *egg albumin* and a *gelatin* solution will be provided. Both are relatively simple proteins and are used because their solubilities and many of their properties are the same as those of soluble serum proteins, enzymes, and polypeptide hormones.

Simple proteins are large polyamide molecules made of α-amino acid monomer units. The amide bonds (peptide links) are hydrolyzed by water, especially in the presence of catalysts such as the peptidase enzymes, acids, and bases. This "digestion" should be thoroughly understood; its chemistry is summarized in the equation:

$$\underset{\underset{G}{|}}{H_2NCH}\overset{\overset{O}{\|}}{C}\text{—}(NH\underset{\underset{G}{|}}{CH}\overset{\overset{O}{\|}}{C})_n\text{—}NH\underset{\underset{G}{|}}{CH}\overset{\overset{O}{\|}}{C}OH + (n+1)H_2O \underset{\text{protein anabolism}}{\overset{\text{catalyst digestion}}{\rightleftharpoons}} (n+2)H_2N\underset{\underset{G}{|}}{CH}\overset{\overset{O}{\|}}{C}\text{—}OH$$

The reverse of this reaction is the simplest way of expressing protein anabolism (the synthesis of protein).

Since n in the formula is commonly of a magnitude of 250 or more, and since a typical protein utilizes some 20 different amino acids (20 different G groups), it should be apparent that the number of isomeric compounds for any specific protein mounts to astronomical numbers. The most incredible thing about protein anabolism is that the host species, for generation after generation, can unerringly synthesize the same specific isomers needed for the various tissue functions. Perhaps the most exciting area of scientific research in this decade has been in molecular biology. The structure and properties of proteins continue to be the subject of intense activity.

MATERIALS NEEDED

All materials, reagents, and reference solutions required will be available in the laboratory.

EXPERIMENTAL PROCEDURE

Try the following tests on the solutions of proteins provided and answer to your own satisfaction the questions posed on the data sheet. It is suggested that for your lab records you list each test tried and write a brief summarizing account of the results observed as well as any notes on chemical or physical reasons for the observed phenomenon. Be brief, and record observations with the idea that you will use your records for future reference.

A. The "Salting Out" Process

To a couple of milliliters of albumin solution in a test tube, add a saturated solution of $(NH_4)_2SO_4$ dropwise and with mixing. If you observe closely, you will see an increase in cloudiness and eventually a separation of tiny protein aggregates as they float to the surface. Do not add a large excess of the salt because it is interesting to reverse the process by diluting with water. Add water with intermittent gentle shaking until all evidence of separated protein has disappeared.

This is one of the few techniques by which a protein can be separated from a solution *without* causing denaturation.

B. The Buffer Activity of the Proteins

To 5 mL of albumin solution add two drops of alizarin red S indicator (yellow, 4.6 to red, 6.0). Then add, dropwise, *counting the drops*, a 0.1 N acetic acid solution until the alizarin red has just turned to yellow. to put the buffer capacity of the protein solution in perspective, try the same acidification of 5 mL of plain water using alizarin red S as the indicator.

The amphoteric behavior of serum albumin accounts for its function as a blood buffer. The pH of the blood must be carefully maintained at about 7.35. Soluble albumins in the blood assist the phosphate and bicarbonate buffers to this end. Were we to select a proper indicator, the buffer capacity of the albumin solution could easily be shown in the other direction by its ability to neutralize a 0.1 N NaOH solution.

C. Alkaloidal Reagents

Certain complex acids such as picric acid, tannic acid, phosphotungstic acid, and phosphomolybdic acid are called alkaloidal reagents. They are capable of forming stable, insoluble salts with the alkaloids. Alkaloids are complex amino compounds that occur naturally. Since the proteins also are complex and have free amino groups scattered about their surface, they too form precipitates with the alkaloidal reagents.

To a couple of milliliters of albumin solution, add a few drops of saturated picric acid solution. What happens? [Picric acid is a good yellow dye for woolens, silk, and linens (protein fabrics). **Don't get it on your skin—it's permanent.**]

D. The Action of Heavy Metals on the Proteins

Metal ions such as Hg^{2+}, Pb^{2+}, Ag^{+}, Cu^{2+}, and Fe^{3+} not only react with carboxylate groups on protein but also coordinate with amino groups. As a result, they are well qualified to precipitate proteins. A rough example will give the general idea:

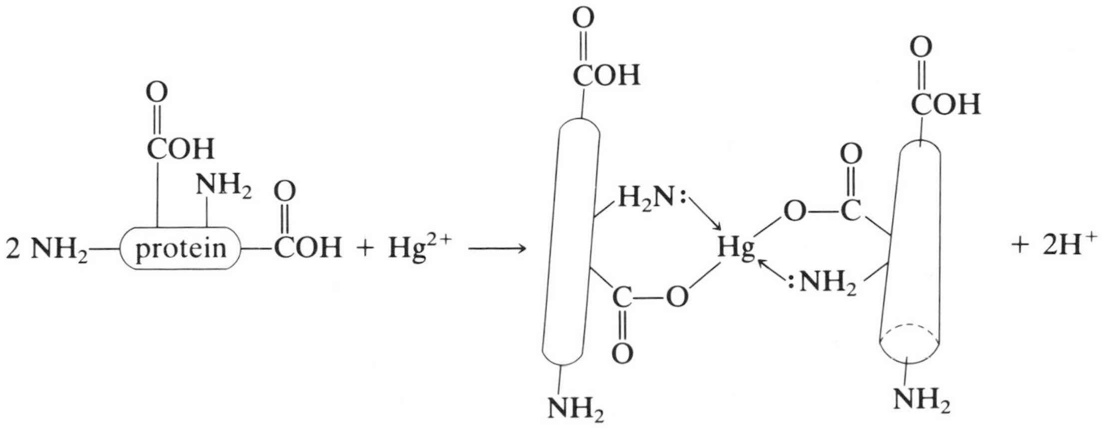

$$2\ NH_2\text{—(protein, with } \overset{O}{\overset{\|}{C}}OH \text{ and } NH_2\text{)—}\overset{O}{\overset{\|}{C}}OH + Hg^{2+} \longrightarrow \text{[two protein chains, each with } \overset{O}{\overset{\|}{C}}OH \text{ and } NH_2 \text{ ends, joined through } Hg \text{ bound to } H_2N\text{:}, \text{:}NH_2 \text{ and two } \text{—}\overset{O}{\overset{\|}{C}}\text{—O— groups]} + 2H^{+}$$

Try adding $HgCl_2$, $AgNO_3$, and $Pb(C_2H_3O_2)_2$ to separate albumin solutions. What happens?

E. The Biuret Test

Place 2 mL of albumin solution in a test tube and make it alkaline with a few drops of NaOH solution. Add a few drops of dilute copper sulfate (1%). Note the color.

Any compound having two or more peptide bonds in close approximation will give this test. Thus any soluble protein can be counted on to give a biuret test.

F. Xanthoproteic Test

To a piece of white wool or silk cloth, add a drop of concentrated HNO_3. Note the color. Rinse with water and then add a couple of drops of ammonia to the colored spot and observe the color change.

The amino acids tyrosine, phenylalanine, and tryptophan have aromatic rings incorporated in their G groups. These (especially tyrosine) react with concentrated HNO_3 to become nitrated. In general, nitro compounds are yellow in color (e.g., picric acid). Nitro compounds form a more intensely colored resonance structure in alkaline solution.

One can readily demonstrate the protein nature of fingernails (they are albuminoids) by *carefully* placing *one* small drop of concentrated HNO_3 on the nail and allowing it to remain there for 10 or 15 seconds. Rinse off with water and note the spot where denaturation took place. Add a drop of ammonia to the spot. Note: This is a fairly permanent spot.

G. Millon's Test

To 3 mL of albumin solution, add a few drops of Millon's reagent. Mix and note what has happened. Carefully heat to near boiling and observe again.

Millon's reagent consists of a solution of the nitrites and nitrates of mercury in concentrated HNO_3. Any protein solution will give a white precipitate when treated with the reagent. Concentrated HNO_3 readily destroys salt bridges and H-bonds and the Hg^{2+} ion will, of course, form complex salts. If, however, the protein contains tyrosine, the active phenol ring is mercurated on heating and the coagulated protein mass changes color and draws together into a colored lump.

Try this same test with gelatin solution and account for any difference in test results.

H. Heller's Ring Test

Place 3 mL of *concentrated* HNO_3 in a dry test tube. Incline the tube as illustrated in Figure 30.1 and slowly and carefully pour a milliliter or two of protein solution down the side of the tube so that a layer is formed on top of the nitric acid. Note the changes occurring at the interface.

The concentrated HNO_3 disrupts existing H-bonds and salt bridges to form nitrate salts with free amine groups. Surprising as it may seem, such

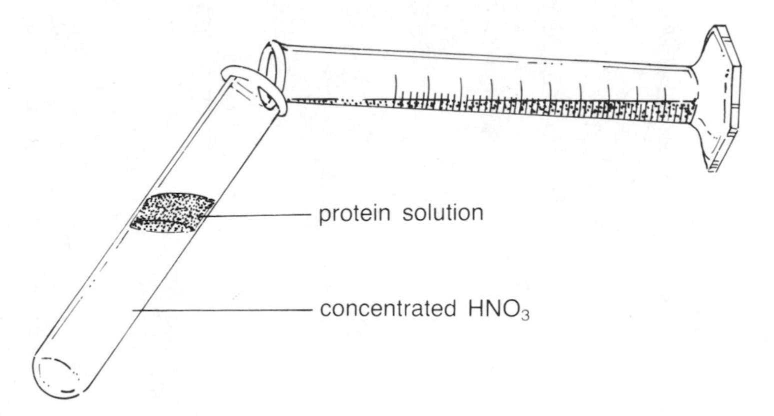

Figure 30.1

salts are generally insoluble. Even urea, $H_2N\underset{\overset{\|}{O}}{C}NH_2$, can be precipitated from solution with concentrated HNO_3 as urea nitrate.

I. Hopkins–Cole Test

In a small test tube mix a milliliter or two of protein solution with an equal volume of Hopkins–Cole reagent. Using the same technique as described for the Heller's ring test, *carefully* run about 5 mL of *concentrated* H_2SO_4 down the side of the tube so there is a minimum of mixing, and a layer of H_2SO_4 forms *under* the protein-reagent layer. A color should develop slowly at the interface provided that the protein being tested contains tryptophan units. Observe the color. A white precipitate or cloudiness at the interface at first indicates a denaturation reaction, but is *not* a positive Hopkins–Cole test. Try the test on both albumin and gelatin solutions.

Hopkins–Cole reagent is essentially an aqueous solution of glyoxylic acid, $\underset{\overset{\|}{O}}{HC}—\underset{\overset{\|}{O}}{C}—OH$.

J. The Coagulation Test

Carefully heat a test tube (filled with the protein solution to the two-thirds level) at a spot about 1 inch below the surface level, as illustrated in Figure 30.2. The heating should be done carefully and without mixing and on one spot only. Note the results and explain what has happened.

When albumin appears in the urine, there is a strong suggestion that a kidney malfunction like nephritis or nephrosis may exist. This heat coagulation test is a simple and reliable means for detecting such albuminuria. In urinalysis a few drops of acetic acid are added and mixed in prior to such tests to dissolve certain phosphates that might also precipitate out on heating and give a false test.

K. The Charring Test for Solid Proteins

Place a *very small* amount of a solid protein such as casein, hair, leather, or feathers in an evaporating dish (or crucible) and heat strongly until it chars. Note the odor.

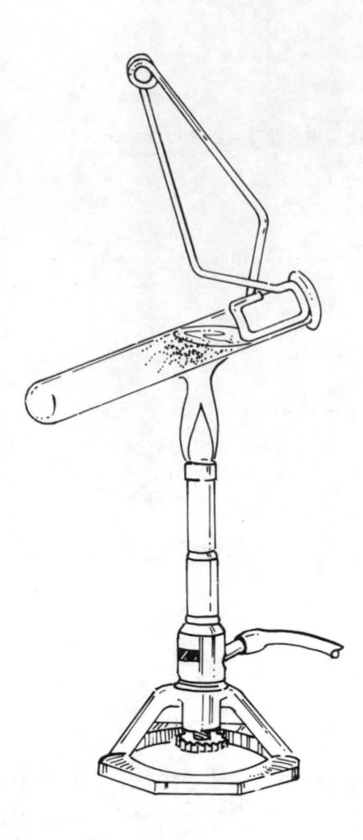

Figure 30.2

Carbohydrates and fats will also char, but the odors are less repulsive since they contain no sulfur.

DATA SHEET

Name: ______________________ Lab Section: ______ Date: __________

THE PHYSICAL AND CHEMICAL PROPERTIES OF PROTEINS

A. The "Salting Out" Process

Explain how a protein is precipitated from solution by thc addition of saturated ammonium sulfate:

B. The Buffer Activity of the Proteins

Complete the indicated neutralization equations for this hypothetical protein:

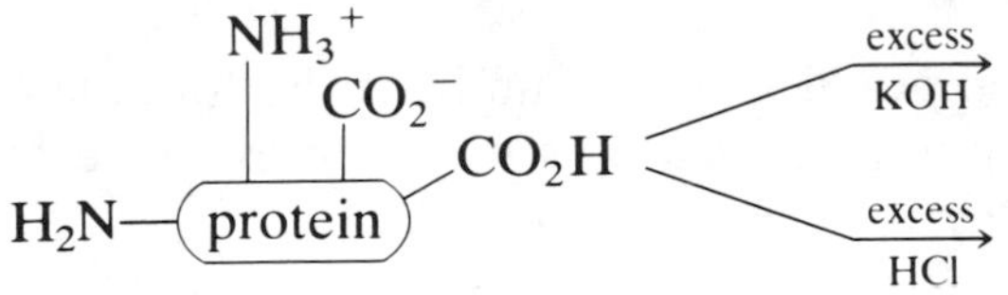

NH_3^+, CO_2^-, CO_2H

H_2N—(protein)

$\xrightarrow[\text{KOH}]{\text{excess}}$

$\xrightarrow[\text{HCl}]{\text{excess}}$

C. Alkaloidal Reagents

a. Complete the reaction for the treatment of the generalized protein structure below with picric acid:

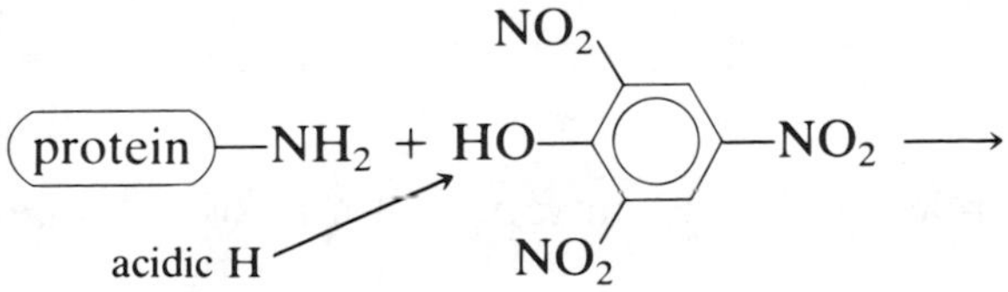

$$(\text{protein})\text{—}NH_2 + HO\text{—}C_6H_2(NO_2)_3 \longrightarrow$$

(Picric acid ring: NO_2 groups at ortho, ortho and para positions; arrow labelled "acidic H" points to HO.)

b. Why is the old-fashioned practice of applying a tea compress to a third-degree burn chemically sound?

c. A certain commercial burn ointment consists of tannic acid and a small amount of phenol in a petrolatum base. List one logical function for each of the three ingredients of the ointment:

(1) Tannic acid: ______________________

(2) Phenol: ________________________________

(3) Petrolatum: ________________________________

D. The Action of Heavy Metals on the Proteins

a. Zinc acetate and zinc chloride solutions have been used as fixatives for histological tissue preparations. How do these heavy metals function in the process?

b. Why is raw egg white so universally recommended as an antidote for poisons like arsenate of lead and paris green (a copper salt)?

E. The Biuret Test

Biuret is a dimeric condensation product of urea and has the formula $H_2N\overset{\overset{\displaystyle O}{\|}}{C}NH\overset{\overset{\displaystyle O}{\|}}{C}NH_2$ (both carbons bearing =O). It gives a beautiful biuret test as might be expected from its name. What structural feature does biuret have in common with the proteins?

F. Xanthoproteic Test

Complete the following reaction for the nitration of a tyrosine moiety on a protein:

$$\text{(protein)}-CH_2-C_6H_4-OH \xrightarrow{\text{concd } HNO_3}$$

G. Millon's Test

Explain the differences observed when applying Millon's test to albumin and gelatin solutions:

H. Heller's Ring Test

Write an equation that illustrates the reaction of concentrated HNO_3 on the zwitterion form of tyrosine (assume an excess of HNO_3):

I. Hopkins–Cole Test

Account for the differing results obtained when albumin and gelatin solutions were subjected to the Hopkins–Cole Test:

J. The Coagulation Test

Suppose that this test is positive when applied to a urine specimen from a patient. What is another symptom the doctor or nurse might look out for? (Refer to Experiment 13.)

EXPERIMENT 31

Starch Digestion: Ptyalin Number

OBJECTIVES

1. To establish and interpret your own ptyalin number.
2. To be able to follow the digestion of starch through the alimentary tract.
3. To be able to design an enzyme study for a hydrolytic enzyme in which optimum pH is established.

PRELABORATORY QUESTIONS

1. Define hydrolysis.
2. What is the chemical nature of ptyalin, and why is the temperature maintained at 37° +/− 2° in this study?
3. Define ptyalin number.
4. What is meant by the "achromic point" watched for in this experiment?

DISCUSSION

In this experiment, some aspects of carbohydrate digestion are investigated. Your own ptyalin number will be established, and a few test reactions learned earlier will be applied. You are encouraged to improve and refine the procedure.

The *ptyalin number* is a quantitative measure of the capacity of a sample of saliva to digest starch. It varies widely from person to person.

Ptyalin Number: The number of milliliters of 1% starch solution that can be digested to the achromic point in 30 minutes by 1 mL of saliva at 37°C.

Ptyalin is the specific starch-splitting enzyme found in saliva. It is protein in nature and subject to the usual denaturation. It is a hydrolytic amylase and catalyzes the partial digestion of starch. The alpha glycoside linkages are cleaved in a stepwise fashion:

$$\underset{\text{Starch}}{(C_6H_{10}O_5)_n} \xrightarrow{xH_2O} \underset{\text{Erythrodextrins}}{x(C_6H_{10}O_5)_{n/x}} \xrightarrow{yH_2O} \underset{\text{Achrodextrins}}{xy(C_6H_{10}O_5)_{n/xy}} \xrightarrow{xy/2\ H_2O} \underset{\text{Maltose}}{n/2\ C_{12}H_{22}O_{11}}$$

The final hydrolytic cleavage doesn't take place until the maltose has reached the small intestine and is hydrolyzed to glucose, $C_6H_{12}O_6$, by the enzyme maltase.

The reagent, iodine in potassium iodide solution (I_2/KI), is the simplest and most effective starch complexing agent. In the presence of starch, a deep blue, heat-sensitive complex develops. The erythrodextrins are somewhat smaller molecules, and their iodine complex is red. The achrodextrins and maltose develop no color in the presence of iodine. They are said to be achromic with respect to iodine.

Pancreatin is a mixture of the three pancreatic enzymes: amylopsin, trypsin, and steapsin. It is extracted from the pancreas of cattle or hogs. All three of its component enzymes exhibit optimum activity in neutral or slightly alkaline environments, such as prevail in the small intestine. In this experiment, the amylopsin is the active enzyme since it is the amylase, and we study the digestion of starch. The trypsin (a peptidase) and the steapsin (a lipase) components of the pancreatin remain idle for lack of suitable substrates.

MATERIALS NEEDED

Everything needed for this experiment is provided in the laboratory.

EXPERIMENTAL PROCEDURE

A. Ptyalin Number

Collect about 10 mL of saliva in a clean test tube. This much you can spare in the interests of scientific investigation since you secrete approximately 1500 mL daily. If it helps, chew on a *small* piece of clean paraffin or dietetic gum to stimulate salivation. If you have eaten recently, it may be best to filter the saliva through a gauze as it is collected.

Prepare saliva solutions of concentrations 66.7%, 33.3%, 10%, 6.7%, 3.3%, 1.0%, 0.7%, and 0.3% as follows: First prepare 20 mL of a 10% saliva solution in a separate, clean test tube by *thoroughly* mixing 2 mL of the collected saliva and 18 mL of distilled water. Similarly, in a beaker of sufficient size, prepare a 1% stock solution of saliva by *thoroughly* mixing 1 mL of pure saliva with 99 mL of distilled water.

Number ten clean test tubes (near the top of the tube) with a wax pencil and prepare the ten solutions as indicated in Table 31.1. Measure the volumes needed with your pipet and use distilled water as the diluent. Mix well but do not put them into your constant temperature bath as yet. This bath, prepared

TABLE 31.1

Tube No.			% Saliva
1	3 mL of pure saliva	:	100.0%
2	2 mL of pure saliva + 1 mL of water:		66.7%
3	1 mL of pure saliva + 2 mL of water:		33.3%
4	3 mL of 10% saliva	:	10.0%
5	2 mL of 10% saliva + 1 mL of water:		6.7%
6	1 mL of 10% saliva + 2 mL of water:		3.3%
7	3 mL of 1% saliva	:	1.0%
8	2 mL of 1% saliva + 1 mL of water :		0.7%
9	1 mL of 1% saliva + 2 mL of water :		0.3%
10	3 mL of distilled water	:	0.0%

ahead of time, should be a beaker of water of such a size that the ten test tubes will stand upright without spilling their contents. A thermometer should be clamped in position so that a constant vigil can be maintained and the temperature kept as close to 37° as possible.

As quickly as possible, add 3 mL of 2% starch solution to each of these ten test tubes, *mix very thoroughly,* and when all have been prepared, place them simultaneously into the water bath. Record the time and regulate the heating so that the bath temperature is maintained 37° +/− 2° for exactly 30 minutes.

Since each tube originally contained 3 mL of saliva or saliva solution, the addition of 3 mL of 2% starch to each will effectively result in 6 mL of 1% starch in each test tube. These 1% starch solutions then will be subject to the digestive action of varying amounts of ptyalin.

After exactly 30 minutes, add just *two drops* of I_2/KI reagent to each test tube and *mix thoroughly.* Record the resultant color on the data sheet. Establish the achromic point (the highest-numbered tube showing no blue or red-violet color). This tube contains the minimum saliva concentration needed to reach an achromic point (only the yellow-brown color of the I_2/KI). Calculate the ptyalin number of your saliva. This can be very easily done by simple proportions. From the definition,

ptyalin number = ______ mL of 1% starch per milliliter of saliva

From Table 31.1, you can see that the possible ptyalin activities are widely separated. If you are interested and have the time, prepare saliva solutions in smaller percent increments between the achromic and the red-violet points indicated in the first rough run. A more exact ptyalin number can then be obtained.

Determine the pH of your saliva using the universal indicator paper or the indicator set provided.

B. Starch Digestion with Pancreatin

Prepare four test tubes as indicated in Table 31.2. To each tube add 2 mL of 2% starch solution; *mix thoroughly;* place the tubes in the water bath, and maintain a temperature of 37° +/− 2° for 30 minutes. At the end of the 30 minute interval, test *half* of each digestion mixture with two drops of I_2/KI solution as before. Record the results on the data sheet and then run a Benedict's test on the remaining half of the solution by adding 4 mL of Benedict's reagent, mixing well, and heating in a boiling water bath (see Experiment 26 for details of the Benedict's test).

Interpret your results by explaining in each case why the iodine test was positive or negative and why the Benedict's test was positive or negative.

The pH in tubes 2 and 3 will approximate 4 and 8, respectively. If in doubt, check with the indicator paper.

TABLE 31.2

Tube #	Composition of Digestion Solution
1	2 mL pancreatin solution plus 2 mL of distilled water
2	2 mL pancreatin solution plus 2 mL of 0.8% HCl solution
3	2 mL pancreatin solution plus 2 mL of 0.5% Na_2CO_3 solution
4	4 mL of distilled H_2O

DATA SHEET

Name: ______________________ Lab Section: ______ Date: __________

STARCH DIGESTION: PTYALIN NUMBER

A. Ptyalin Number

Volume of samples used: ______ Temperature range: __________

Reaction time: ______

Tube Number	Percent Saliva	Color Reaction with Iodine
1	______	______________
2	______	______________
3	______	______________
4	______	______________
5	______	______________
6	______	______________
7	______	______________
8	______	______________
9	______	______________
10	______	______________

The percent saliva in the achromic sample is: ______________

The ptyalin number of my sample is (give a range): ______________

B. Starch Digestion with Pancreatin

Digestion time: ____________ Digestion temperature range: ____________

Tube No.	Approx. pH	Results of Iodine Test	Results of Benedict's Test
1	________	____________	____________
2	________	____________	____________
3	________	____________	____________
4	________	____________	____________

The optimum pH for the digestive action of pancreatin on starch is: ____________

Interpretation and rationalization of the results obtained:

QUESTIONS

1. In the discussion, the digestions of starch, dextrins, and maltose were localized and accounted for. What happens to sucrose, lactose, and cellulose? (Refer to your textbook and indicate what happens and where it occurs.)

   a. Sucrose

   b. Lactose

   c. Cellulose

2. Ptyalin, also called salivary amylase, is classified as an α-1,4-glucosidase. Explain the significance and meaning of this classification.

3. In tube 3 of the pancreatin investigation, you mixed 2 mL of 2% pancreatin, 2 mL of 0.5% Na_2CO_3, and 2 mL of 2% starch solution.

   a. What was the approximate percent by weight of Na_2CO_3 in the reaction tube?

b. What was the molarity, M, of the Na_2CO_3 in the reaction tube?

4. It can be readily calculated that the HCl concentration in tube 2 of the pancreatin investigation was 0.074 M. Ignoring the buffer capacity of the protein pancreatin, what is the approximate pH of this reaction mixture?

Fat and Protein Digestion

OBJECTIVES

1. To be able to trace the digestion of proteins and fats.
2. To learn the optimum conditions for gastric and intestinal digestion.
3. To demonstrate the functions of bile.
4. To learn the composition of bile and chemical tests for bile salts and cholesterol.

PRELABORATORY QUESTIONS

1. Where is most of the ingested fat digested, and which enzyme is responsible for this digestion?
2. What is a metabolite?
3. The digestions studied in this experiment are carefully controlled to maintain a temperature between 36° and 39°. Why?
4. How are the bile salts and ordinary soap alike?

DISCUSSION

This experiment is an investigation of enzymatic reactions and the conditions that promote optimum enzyme activity. Egg albumin, a vegetable oil, and milk are digested in vitro. You are invited to extend and refine the investigation. You will probably need to review carefully the textbook material on fat and protein digestion in order to complete your data sheet.

Fats and proteins undergo little if any salivary digestion, but as the masticated food mass accumulates in the stomach, chemical changes begin. Stomach acid (HCl) is secreted in the gastric mucosa, and in turn pepsinogen and gastric lipase are activated. The pepsin released proceeds with the partial hydrolysis of proteins while the gastric lipase goes to work on those fats that happen to be already emulsified. The pH for this gastric digestion usually lies between 1 and 3.

The final touches in the digestion process require different conditions. The acidic food mass (chyme) is passed through the pyloric valve into the duodenum. Here its intrinsic acidity stimulates the gallbladder to release bile—an alkaline solution of bile salts—and pancreatic enzymes. The bile salts emulsify fats and make them more vulnerable to enzyme-catalyzed digestion. With the aid of peristaltic action, the chyme is neutralized, the action of pepsin stops, fats are emulsified, enzymes from the pancreas and the intestinal mucosa are activated, and a whole new digestive environment is created. The pH settles at around 8. Steapsin hydrolyzes the now emulsified fats while trypsin, chymotrypsin, aminopolypeptidase, carboxypolypeptidase, and dipeptidase finish the job on the proteins. The fats and proteins are thus hydrolyzed to absorbable metabolites.

MATERIALS NEEDED

All solutions and necessary materials are provided in the laboratory except a mortar and pestle, which you can check out if the cholesterol is provided in the form of gallstones.

EXPERIMENTAL PROCEDURE

In this experiment you will run some test tube digestions and a few miscellaneous tests on bile. To save laboratory time, work in pairs and set up a large beaker for a water bath, as in Experiment 31.

A. Digestion Mixtures

At the beginning of the period, number 12 clean 10 mL test tubes near the top with a wax pencil and prepare the digestion mixtures indicated in Table 32.1. Mix all tubes thoroughly and keep them in the water bath at 36–39° until about 15 minutes remain in the period. Then study each tube and record your results. In your recorded observations also indicate briefly why a reaction did or did not occur. In the hour or so during which the digestions for part A are proceeding, part B can easily be run.

B. Tests for Bile Salts, Cholesterol, and Proteins

Gmelin's Test for Bilirubin In cases of jaundice resulting from an obstruction of the bile duct, bile is excreted in the urine. Gmelin's test is often used to detect the presence of bile in urine. It depends on the colored oxidation products that result when reduced bile pigments are oxidized by concentrated HNO_3.

Place 5 mL of concentrated HNO_3 in a test tube. Holding the test tube at a slant (as shown in Figure 30.1), carefully add 3 mL of water containing a drop or two of fresh bile so that a layer is formed above the HNO_3.

Record your observations on the data sheet.

TABLE 32.1

Tube #	Reaction Mixture	Tube #	Reaction Mixture
1	*Very small thin slice* of egg white plus 6 mL of distilled water	7	6 mL of whole milk plus 1 mL of litmus solution plus 3 mL of distilled water plus 1 mL of 0.5% Na_2CO_3
2	*Very small thin slice* of egg white plus 3 mL of 2% pepsin solution plus 3 mL of distilled water	8	3 mL of whole milk plus 1/2 mL of litmus solution plus 2 mL of pancreatin solution plus 1/2 mL of 0.5% Na_2CO_3 solution
3	*Very small thin slice* of egg white plus 3 mL of 2% pepsin solution plus 3 mL of 0.8% HCl solution	9	1 *very small drop* of safflower oil* plus 6 mL of distilled water
4	*Very small thin slice* of egg white plus 3 mL of 2% pepsin solution plus 3 mL of 0.5% Na_2CO_3 solution	10	1 *very small drop* of safflower oil* plus 5 mL of distilled water plus 1 mL of a solution of bile salts (or of fresh bile)
5	*Very small thin slice* of egg white plus 3 mL of 2% pancreatin solution plus 3 mL of 0.5% Na_2CO_3 solution	11	1 *very small drop* of safflower oil* plus 3 mL of steapsin solution plus 2 mL of distilled water plus 1 mL of a solution of bile salts (or of fresh bile)
6	6 mL of whole milk plus 6 drops of rennin solution	12	1 *very small drop* of safflower oil* plus 3 mL of steapsin solution plus 3 mL of distilled water

* (or other vegetable oil)

Hay's Test for Bile Salts The surface tension of water is lowered by the presence of surface-active agents such as soap and bile salts. This phenomenon is utilized in the Hay's test for bile in urine.

Place 6 mL of distilled water in each of three test tubes. To the first add two drops of soap solution, and to the second add two drops of a fresh bile solution. Mix thoroughly, and into each test tube dust a *very, very small* pinch of powdered sulfur. You need *only* enough sulfur to detect it on the liquid surface. Tap the tubes gently and watch the floating sulfur.

Describe and explain your results on the data sheet.

The Lieberman–Burchard Test for Cholesterol Cholesterol is very soluble in ether. Thus this lipid in a tissue can be detected by first extracting it from the tissue with ether, evaporating off the ether, and then testing the resulting (dried) crystals as described below.

Dissolve a single, tiny crystal of cholesterol in 2 mL of dry chloroform in a clean, dry test tube. Add six drops of acetic anhydride (about 0.5 cc) and two drops of concentrated H_2SO_4 and mix thoroughly. Watch for the color changes and record your results. Test also the $CHCl_3$ extract of gallstones provided. (*Absolutely dry conditions are imperative. Be sure your dropping pipets and test tubes are perfectly dry. Even then, the striking color changes are transitory and fleeting. When tissues are extracted to obtain cholesterol, the ether extract must be thoroughly dried before the tests are made.*)

The Bromine/Freon 113/ Test for Unsaturation Dissolve a couple of crystals of cholesterol in 1 mL of freon 113 in a dry test tube. To this solution, add a few drops of Br_2/freon 113 and observe. Blow a puff of moist air over the mouth of the test tube and watch for a possible HBr–H_2O fog. Record and interpret what you have observed.

The Biuret Test on Enzymes Test your saliva as well as the 2% pancreatin solution provided for protein character by mixing a small volume (a milliliter or less) with an equal volume of 10% NaOH and then adding three or four drops of biuret reagent (0.5% $CuSO_4$).

DATA SHEET

Name: ______________________________ Lab Section: ______ Date: ____________

FAT AND PROTEIN DIGESTION

A. Digestion Mixtures

Digestion time: ________________ Temperature range: ________________

Tube Number	Substrate	Enzyme	Approx pH	Observations
1				
2				
3				
4				
5				
6				
7				
8				
9				
10				
11				
12				

B. Tests for Bile Salts, Cholesterol, and Proteins

Gmelin's test Observations:

Hay's test Observations and explanation:

Lieberman–Burchard test Observations:

Bromine/freon 113 test Observations and interpretation:

Biuret test

Observations on saliva:

Observations on pancreatin solution:

QUESTIONS

1. Egg albumin was *best* digested under the conditions of tube number ________, where the pH was approximately ________ and the conditions most closely approached those of the (mouth, stomach, small intestine) ________________________________.

2. The purpose for tubes 1 and 9 was:

3. Using the generalized formula for a simple protein, write an equation for the digestion of a protein in the stomach.

4. The enzyme in pancreatin that is specific for the hydrolysis of proteins is ________________ and it functions best at a pH of ________.

5. The normal pH range for the stomach is ________________________.

6. Write a word equation for the phenomenon observed in tube 6.

7. Write an equation illustrating the complete digestion of glyceryl tributyrate as it might occur in the duodenum of the small intestine.

8. Explain the color change observed in tube 8.

9. Safflower oil (or vegetable oil) was most efficiently digested in tube number ________, where the pH was ________ and the conditions approximated those of the ________________ of the alimentary tract.

10. From observations of tubes 9 and 10, what functions do the bile salts have in intestinal digestion?

11. In Gmelin's test, what are the name and color of the oxidation product of bilirubin?

12. Two of the common bile salts are ________________________ and ________________________.

13. Three places in the human body where cholesterol is commonly found are ____________, ____________, and ____________.

14. Complete the equation for the acid-catalyzed reaction between cholesterol and acetic anhydride. (See Experiment 22 if this esterification is new to you.)

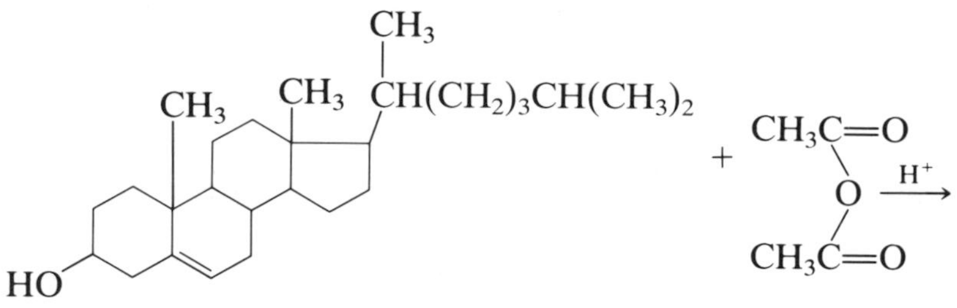

CH_3

CH_3 CH_3 $CH(CH_2)_3CH(CH_3)_2$

HO

$+$ $CH_3C{=}O$ — O — $CH_3C{=}O$ $\xrightarrow{H^+}$

15. From its functional groups, cholesterol belongs to what classes of organic compounds?

The Chemical and Physical Properties of Blood

OBJECTIVES

1. To understand the difference between serum, plasma, and formed elements, and to know how these fractions are isolated.
2. To understand the effect of osmotic changes on blood cells.
3. To learn how to identify and remove blood stains.
4. To learn how to measure the buffer capacity of blood.

PRELABORATORY QUESTIONS

1. What is oxalated whole blood and why is it so treated?
2. What is an isotonic solution?
3. When whole blood is centrifuged, what is removed and what is the remaining solution called?
4. What is the main buffer system of the blood?
5. How do whole blood, blood plasma, and blood serum differ?

DISCUSSION

You have learned from your textbook that blood is indeed a very complex solution. Its solutes range from small ions to huge polymers, from true solutes through suspended colloids to cells that exceed colloidal dimensions. It is no wonder that blood is highly sensitive to even small changes in its environment.

When *whole blood* is centrifuged, its cellular components settle out. These are called the *formed elements* and consist of the red corpuscles (erythrocytes), the white cells (leukocytes), and the platelets (thrombocytes). The pale yellow fluid remaining is called *blood plasma*. A great many common blood analyses are made on the plasma alone, for it contains the same metabolytes, wastes, and colloidal material that the whole blood possessed prior to centrifugation.

If freshly collected whole blood is stirred vigorously with a wooden paddle, the fibrinogen is converted to fibrin, which in turn sticks to the paddle and can be removed. The resulting *defibrinated whole blood* is no longer capable of clotting but it does still retain its formed elements.

When whole blood is allowed to stand quietly, a soft clot forms. As the fibrin fibers intermesh as a sort of brush heap, they occlude the filterable formed elements. In time the clot contracts, holding the formed elements and squeezing out a pale yellow fluid called *serum*. Serum is, then, simply whole blood from which fibrinogen and formed elements have been removed. It still contains a complete complement of albumins and globulins, which are appropriately called serum proteins.

Since whole blood begins to clot immediately after being exposed to air, it must be treated in some way to prevent its clotting if it is to be saved and either reused or analyzed. Since the clotting process involves a series of chemical reactions and since one of the essential reactants in the series is calcium ion, the removal of Ca^{2+} is the simplest way to prevent clotting.

Oxalated whole blood is blood that has been treated immediately upon collection with an isotonic solution of sodium oxalate. As you may recall from general chemistry, calcium oxalate is insoluble in water and precipitates out quantitatively. In your study of blood under the microscope in this experiment, you may well spot some of the beautiful microcrystals of calcium oxalate. Oxalates, however, are poisonous, so blood destined for intravenous use must be treated in a safer way.

Citrated whole blood is blood that is treated immediately upon collection with an isotonic citrate solution. The citrate ion has the capacity for tying up (chelating) calcium ion. It does not precipitate the Ca^{2+} out as does oxalate, but it does remove it as a possible active participant in the clotting process. Citrate is a totally acceptable metabolite (the citric acid cycle), and so citrated whole blood is perfectly safe for intravenous use.

In this experiment oxalated whole blood will be provided. You will study it as well as plasma prepared from it to learn a few common chemical and physical properties. Do not be alarmed if your prepared plasma takes on more of a pink than a yellow color. Hemolysis is difficult to prevent. Take care nonetheless to use dry flasks and test tubes, and hemolysis will be minimized.

MATERIALS NEEDED

Obtain a buret and holder. Reagents, microscopes, centrifuges, oxalated whole blood, and other necessary materials are available in the laboratory.

EXPERIMENTAL PROCEDURE

Run the tests on the blood plasma in the order presented, and report your observations clearly and in detail on the data sheet.

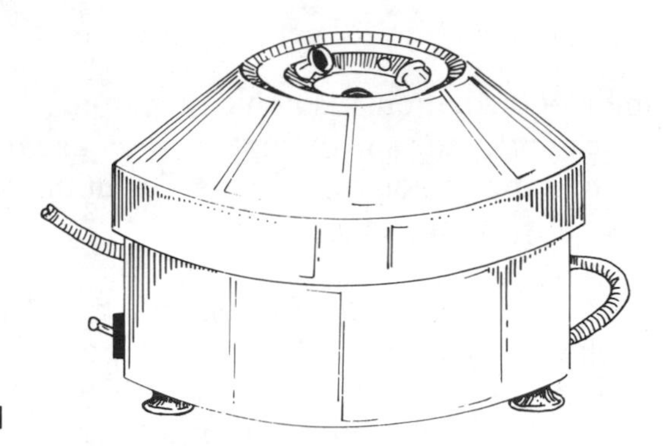

Figure 33.1

A. Blood Plasma

Select two test tubes of equivalent size and of such dimensions that they fit nicely in opposite wells of the centrifuge (Figure 33.1). Fill them to an equal depth with oxalated whole blood, place in diametrically opposite wells in the centrifuge, and run at full speed for enough time to get a clear deposition of the red formed element fraction in the bottom of the tube (5–10 minutes). In this manner, collect at least 12 mL of the pale yellow plasma for use in the next four tests.

Blood plasma contains amphoteric proteins and amino acids. As a result, the determination of buffer capacity is complicated by the buffering action of these proteins. End points are not precise. Note also that in the test for glucose, a coagulation precipitate of fibrinogen with the Cu^{2+} interferes with the test interpretation.

Blood Sugar Heat 3 mL of Benedict's reagent in a test tube in a water bath, as described in Experiments 26 and 31. When the temperature is approximately 100°, add $\frac{1}{2}$ mL of plasma, mix well, and continue heating. On the data sheet, note and record, and explain what happens.

Plasma Protein To 1 mL of plasma, add a few drops of Millon's reagent and heat slowly to boiling (see Experiment 30). Record and explain your observations.

Buffer Capacity for Acids (meq/liter) Rinse and fill a buret with the 0.05 N HCl provided. In a 125 mL Erlenmeyer flask place exactly 5.0 mL of plasma, 15–20 mL of distilled water, and a few drops of Congo red indicator. Titrate to the blue end point with the HCl. (Sometimes, due to partial hemolysis prior to centrifugation, the plasma will have a pink color to start with. In that case, the end point is more of a mud color, but it is detectable.) Report the buffer capacity for acid in terms of milliequivalents of acid per liter of blood.

Buffer Capacity for Bases (meq/liter) Rinse and fill a buret with the 0.05 N NaOH solution provided, and use this to titrate 5 mL of plasma diluted with 15–20 mL of water. Use as an indicator either metacresol purple (yellow to purple; pH 7.4–9.0) or thymol blue (yellow to blue; pH 8.0–9.5). Calculate the buffer capacity for base in terms of milliequivalents of base per liter of blood.

B. Whole Blood

Prepare the following blood solutions (isotonic, hypotonic, and hypertonic) as directed, but study each under the microscope *as soon as possible after mixing*; i.e., do one at a time as you make the slides in order to actually observe the osmotic effects as they take place.

Isotonic Blood Solution Add $\frac{1}{2}$ mL of whole blood to 1 mL of physiological saline solution (0.85% NaCl). Mix and prepare a microscope slide. For your data sheet, diagram the red cells as they appear under the high power. Set the rest of the solution aside for a later comparative inspection.

Hypotonic Blood Solution Add 1 drop of whole blood to a few drops of 0.5% NaCl on a microscope slide. Quickly cover and observe; describe and sketch the phenomenon on your data sheet.

Hypertonic Blood Solution Add a few drops of whole blood to 2 mL of a 2% NaCl solution. Quickly prepare a microscope slide and observe the phenomenon. Describe, explain, and draw the phenomenon on your data sheet.

Comparison of the Physical Appearances of Isotonic, Hypotonic, and Hypertonic Blood Solutions Place 1 mL of oxalated whole blood in each of three dry test tubes. To the first add 2 mL of physiological saline solution (0.85% NaCl). To the second add 2 mL of water, and to the third, 2 mL of 2% NaCl. Make notes on the differences in their appearance.

Centrifuge all three tubes and again observe for differences. Record all observations on your data sheet.

C. Tests for Blood and Removal of Blood Stains

The hemoglobin in blood is a peroxidase. That is, it catalyzes the decomposition of hydrogen peroxide to water and nascent oxygen:

$$H_2O_2 \xrightarrow{\text{peroxidase}} H_2O + [O] \longrightarrow \tfrac{1}{2}\, O_2 \uparrow$$

A powerful oxidant, atomic oxygen will convert colorless TMB to a deep blue-green oxidation product, something ordinary O_2 does not readily accomplish. Herein lies the secret of the TMB test for blood. Extremely sensitive, it works on even trace amounts of dried blood that may be years old. It is a favorite test among forensic chemists.

TMB (3, 3′, 5, 5′-Tetramethylbenzidine) is a colorless, noncarcinogenic substitute for benzidine. It is oxidized to a blue-green quinonimine by atomic oxygen.

Action of a Peroxidase on H_2O_2 Place a drop of blood on the cleaned bottom of an upturned beaker and add a drop of 3% peroxide. Record your observation on the data sheet and write an equation for what has happened.

TMB Test on Dilute Blood Solutions Place a drop of blood in a test tube filled with water. Empty and, without rinsing, fill once again with water. This should be an essentially colorless and very dilute blood solution.

To this very dilute blood solution, add 1 drop of TMB reagent and then one drop of 3% hydrogen peroxide, and mix. A blue color is evidence of the presence of a peroxidase (hemoglobin).

TMB Test on Blood Stains Add a drop of TMB reagent to a slip of cloth (provided) with an old, dried blood stain on it. Next add a drop of 3% H_2O_2. Note the results on the data sheet.

Removal of Blood Stains Rinse the blood-stained spot on a piece of fabric first with cold water. Add a few drops of 3% H_2O_2 to the spot and gently work it into the stain. Rinse thoroughly with cold water. This is a practical hospital and household technique, but it should never be used on colored fabrics.

DATA SHEET

Name: ______________________________ Lab Section: ______ Date: __________

THE CHEMICAL AND PHYSICAL PROPERTIES OF BLOOD

A. Blood Plasma

1. Results of Benedict's test: ______________________________

____

2. Results of Millon's test: ______________________________

What component in plasma protein accounts for the positive test? __________

____

3. Buffer capacity for acids (meq/liter): ______________________

4. Buffer capacity for bases (meq/liter): ______________________

B. Whole Blood

1. Diagram of erythrocyte in isotonic solution:

2. Description and sketch of erythrocyte in hypotonic solution:

3. Diagram of erythrocyte in hypertonic solution:

4. Describe the differences observed among the three solutions and explain why they occur:

C. Tests for Blood and Removal of Blood Stains

1. Observations on the action of a peroxidase on H_2O_2: ____________
   The balanced equation for the action of peroxidase on H_2O_2:

2. Observations on use of the TMB test on blood in solution: ____________

   ____

3. Observations on use of the TMB test on a blood stain: ____________

4. Stain removal: Why should this method of removing blood stains be avoided when colored fabrics are involved? ____________

   ____

QUESTIONS

1. Physiological saline solution (0.85% NaCl by weight) is isotonic to whole blood. Using van't Hoff's equation (Experiment 13), calculate the osmotic pressure of whole blood at body temperature (37° C) in atmospheres. (Remember that NaCl dissociates 100% into Na^+ and Cl^- ions.)

2. The H_2CO_3/HCO_3^- system is the main buffer of the blood. Write equations to show what happens when:

   a. HCl is introduced into this buffer system:

   b. KOH is introduced into this buffer system:

Urinalysis

OBJECTIVES

1. To become familiar with the normal and abnormal constituents of urine covered in this experiment.
2. To know some of the causes for the presence of abnormal constituents in the urine.

PRELABORATORY QUESTIONS

1. To what is the straw yellow color of normal urine attributed?
2. When albumin is detected in the urine (albuminuria), what pathological condition is suggested?
3. How does renal diabetes differ from diabetes mellitus?
4. Write the structure for (a) acetone, (b) acetoacetic acid, and (c) beta-hydroxybutyric acid.
5. When a urine specimen is dehydrated, a residue of solids remains. List two inorganic salts one might expect to find in this residue.

DISCUSSION

A great many diagnostic facts can be obtained from a thorough analysis of a urine specimen. Any deviation from well-established norms can be significant, and the trained specialist, given also a reliable case history, can predict probable causes for the deviations. A brief background in theory is presented with each test. You are strongly encouraged to study your textbook material on urine before embarking on this experiment.

MATERIALS NEEDED

Check out a urinometer and an unknown pathological urine specimen. All other necessary reagents and reference compounds will be available in the laboratory.

EXPERIMENTAL PROCEDURE

Nine tests and observations are described for this urinalysis. Run each test on three parallel specimens: your unknown, your own specimen (which will be specified as the normal specimen on the data sheet), and the pathological control recommended in the procedure. Record all observations and conclusions in detail on your data sheet for use as a reference in your tentative diagnosis.

In preparing for this experiment, each student should collect a urine specimen of his or her own to be tested along with the unknown provided. A 24 hour specimen collected after the first voiding in the morning through and including the first voiding of the following morning would be the most significant. For a 24 hour specimen, the collection flask should contain about 3 mL of toluene, and after each collection the sample should be shaken with this toluene. In this way bacterial action is inhibited and the sample is better preserved.

A. Color

Normal urine has a straw yellow color. The color is deeper in cases of oliguria and lighter in case of polyuria. Compare your sample(s) with pathological specimen 4, which contains bile, and with 1, which is diluted with water. Also note the color of 5, which contains blood. Do your specimens appear abnormal in any way? If so, give a possible reason for the abnormality observed.

B. Specific Gravity

Obtain a urinometer (hydrometer) and determine the specific gravity of your unknown as well as that of pathological specimen 1. The hydrometer operates on the principle of Archimedes: A floating body displaces its own weight of liquid. The urinometer has been calibrated so that the specific gravity can be read directly on the stem at that point where the stem breaks the surface of the liquid.

Fill the cylinder two-thirds full of urine. Remove any foam or floating toluene with a strip of paper toweling, and carefully float the urinometer in the cylinder in such a way that it makes no contact with the edges of the cylinder (see Figure 34.1). (Give it a spin as you insert it.) Normal urine has a specific gravity range of 1.008 to 1.030.

Long's coefficient (2.6) is used to obtain an approximate figure for the total solids in a urine specimen. This coefficient, 2.6, multiplied by the significant figures from the fractional part of the specific gravity of the sample gives the approximate number of grams of total solids per liter of the urine specimen. An example can best illustrate this calculation. Suppose that the

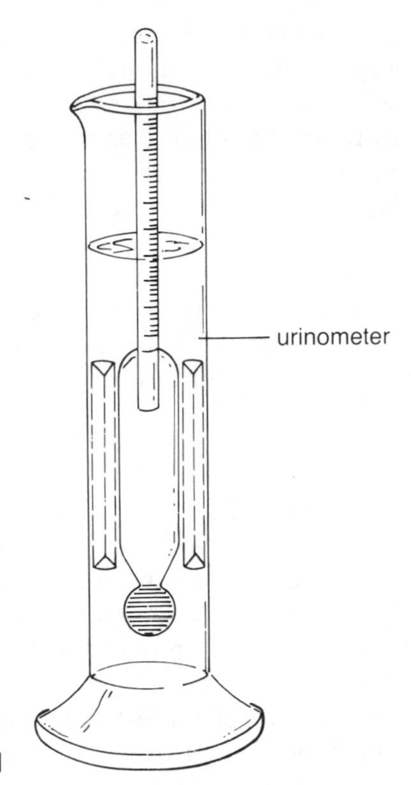

Figure 34.1

specific gravity of a urine specimen is 1.022. The total solids in grams per liter would then be about 22 × 2.6 or 57 g/liter. Normal urine usually has 40 to 60 grams of total solids per liter. In other words, 1 liter of this urine, when evaporated carefully to dryness, would leave a residue of about 57 g.

Determine the approximate total solids per liter of your sample(s), and interpret specific gravity findings. For comparison, you might try also pathological specimen 1, which is typical of the urine from one suffering from diabetes insipidus.

C. pH

Normally, urine has a pH between 5.5 and 7.0, but this can drop lower (during acidosis). Determine the pH of your sample(s) with whatever materials are provided. A comparison should be made with the pH of pathological specimen 2, which is typical of a urine specimen taken shortly after the ingestion of a meal heavy in fruits and vegetables.

A rough pH determination can be made using Nitrazine paper. A drop of the urine touched to the paper will give a yellow color if the pH is about 5, olive at about 6, gray-blue at about 7, and blue at about 8. You may otherwise use a pH meter or an indicator set.

D. Albuminuria

When kidneys are not functioning properly (nephritis or nephrosis), serum albumin can escape into the urine. This loss of albumin from the blood can appreciably decrease the serum osmotic pressure, and water diffuses into the tissues (edema).

We will use the simple heat coagulation test (see Experiment 30 on proteins) to detect the presence of albumin, although Heller's ring test could be used as well.

A tube three-fourths full of urine is held at a slant and the upper third is heated carefully with a burner flame (see Figure 30.2). If a cloudiness develops and remains even after three to four drops of dilute acetic acid have been added, the presence of albumin is indicated. The acetic acid will dissolve phosphates, which often precipitate during the heating and give a false test. Run this test on your unknown(s) and on pathological specimen 3, which contains albumin. If your specimen indicates albuminuria, explain how it might be accounted for, pathologically and otherwise.

E. Glycosuria

Reducing sugars and uric acid give a positive Benedict's test. Uric acid, however, is normally present in such small quantities that its contribution to the reduction of the Cu^{2+} is negligible. In cases where uric acid is present in large amounts, Nylander's reagent (a bismuth complex) can be used in place of Benedict's reagent. The bismuth ion is not reduced to free bismuth by uric acid, but is so reduced by glucose.

Place 5 mL of Benedict's reagent in a clean test tube and heat in a boiling water bath. When hot, add $\frac{1}{2}$ mL (about eight drops) of urine, *mix thoroughly*, and heat in the water bath for 5 minutes more. A rough approximation of the percent sugar can be made from the color that results: green, about $\frac{1}{2}$%; olive green, about $\frac{3}{4}$%; yellow, 1%; and orange-red, 2% or more. Run this test on your specimen(s) and on pathological specimen 1, which contains glucose.

F. Ketonuria

Of the ketone bodies, both acetone and acetoacetic acid can easily be detected. We shall use the Lange test, which detects the presence of acetone only. It is, of course, a fairly safe assumption that if acetone is present, the other ketone bodies will be present also.

Place 1 mL of urine in a test tube and add one drop of glacial acetic acid and *one tiny* crystal of sodium nitroprusside. Shake until the crystal has dissolved. Hold the test tube in an inclined position and carefully layer about 1 mL of concentrated ammonia over the urine solution. A red-to-purple ring at the interface indicates the presence of acetone. Normal urine will give a brown ring. Creatinine, a normal constituent of the urine, will give a red color and interfere with the test unless the acetic acid is added. Run this test on your unknown(s) and on pathological specimen 1, which has a small amount of acetone in it. Interpret your results. (*Note*: The colored ring develops slowly.)

G. Blood

The TMB reagent must be prepared shortly before it is to be used. To prepare the reagent, mix about $\frac{1}{2}$ mL of the TMB–glacial acetic acid solution in a clean test tube with about $\frac{1}{2}$ mL of 3% hydrogen peroxide solution. (TMB is discussed in more detail in Experiment 33.)

A urine specimen containing any traces of hemoglobin will develop a green to blue color when mixed with TMB reagent. For some reason, possibly interference from other urine constituents, the test often fails when the TMB is added to the urine but works when urine is added dropwise to the TMB reagent. For this reason, do the test the second way: Add urine to TMB. Test your unknown, your normal specimen, and control 5. Record your observations.

H. Bile

If bile has found its way into the urine, bile salts will be there also. The bile salts are surface-active agents and will reduce the surface tension of a urine specimen to a point where fine sulfur powder will not float on its surface. This is Hay's test, introduced in Experiment 32.

Dust a very small pinch of dry flowers of sulfur on the surface of a few milliliters of the urine to be tested. Gently tap the tube and observe whether the sulfur particles begin to slowly sink. Test your samples and compare the results with those obtained when pathological specimen 4, which contains bile, is similarly tested.

When human urine containing bile is shaken, a yellow-green foam should develop. Try this on your samples and on pathological specimen 4. Interpret your findings in your report. (*Note*: Pork and beef bile produce foam but only a very pale yellow coloration.)

I. Urea

The average adult excretes about 30 g of urea per day. Urea is the principal end product of amino acid catabolism. Add about a milliliter of urease solution to 5 mL of a urine specimen and allow to stand at about 37° for 10 minutes. Add 1 mL of 20% NaOH and note the odor as compared to that of the original specimen. Write the equation for the reaction that takes place (if indeed it does) and interpret your findings.

Throughout this experiment please leave the water running gently in the sinks. Be certain that the hydrometers and cylinders are clean before returning them to the storeroom. Rinse all leftover reagents carefully down the drain and do not spill on the desks or on yourself! Also, please dispose of any remaining urine specimens immediately after completion of this experiment.

DATA SHEET

Name: ______________________ Lab Section: ______ Date: __________

URINALYSIS

Unknown number __________ contains ______________________

____

Preliminary diagnosis:

OBSERVATIONS AND TEST RESULTS

A. Color

Unknown Number: __________ ______________________

Normal Specimen: ______________________

Control 4: ______________________

Control 1: ______________________

Conclusions:

B. Specific Gravity

Unknown Number: __________ ______________________

Normal Specimen: ______________________

Control 1: ______________________

Conclusions:

C. pH

Unknown Number: __________ ______________________

Normal Specimen: ______________________

Control 2: ______________________

Conclusions:

D. Albuminuria

Unknown Number: __________ ______________________________

Normal Specimen: ______________________________

Control 3: ______________________________

Conclusions:

E. Glycosuria

Unknown Number: __________ ______________________________

Normal Specimen: ______________________________

Control 1: ______________________________

Conclusions:

F. Ketonuria

Unknown Number: __________ ______________________________

Normal Specimen: ______________________________

Control 1: ______________________________

Conclusions:

G. Blood

Unknown Number: __________ ______________________________

Normal Specimen: ______________________________

Control 5: ______________________________

Conclusions:

H. Bile

Unknown Number: __________ ______________________________

Normal Specimen: ______________________________

Control 4: ______________________________

Conclusions:

I. Urea

Unknown Number: __________ ______________________________

Normal Specimen: ______________________________

Conclusions:

QUESTIONS

1. A specimen of urine gave a strong and positive Benedict's test. What additional test(s) might you employ in order to establish whether the glycosuria was indeed due to diabetes mellitus?

2. In one treatment for epilepsy in children, a high-fat diet is employed. The treated children often develop ketosis. Why?

3. What is the chemical origin of the acetone found in the urine of patients having ketosis?

4. Write an equation for the hydrolysis of urea by the action of urease.

QUESTIONS

1. [illegible]

2. [illegible] Why?

3. [illegible]

4. [illegible]

Bromsulphalein Analysis: A Liver Function Test

OBJECTIVES

1. To learn how to use a spectrophotometer.
2. To be able to construct and use a colorimetric reference curve.
3. To gain a comfortable working acquaintance with Beer's law.
4. To understand a simple method for evaluating liver function.

PRELABORATORY QUESTIONS

1. From your text background, list three important functions of the liver.
2. Bromsulphalein is injected for liver function determinations in the dosage 5 mg/kg of body weight. If a patient weighs 172 lb, how many grams of bromsulphalein are needed?
3. The bromsulphalein system conforms to Beer's law. What does this statement mean?
4. A serum specimen must inevitably contain some yellow-colored urobilin from hemoglobin degradation in the liver. This colored bile pigment interferes to a certain extent with the colorimetric determination for bromsulphalein. Why can't we simply pass our serum specimen through a disposable extraction column for clarification as was done on the specimen analyzed for glucose in Experiment 27?

DISCUSSION

How's your liver today? Though jocular and impudent, this greeting gets right to the point. One can feel no better than one's liver permits. Health and well-being are intimately associated with liver function. It is the liver that processes what the blood brings to it—utilizing the metabolites, minerals, and vitamins, and detoxifying and rejecting wastes and potentially harmful chemicals for disposal via the final clearinghouse, the kidneys. The efficiency with which this purifying process proceeds is called "liver function." In this experiment you will apply a common liver function test on a hypothetical serum specimen and diagnose the situation represented.

In the clinical setting, a patient first gives 10 mL of blood. The serum from this first specimen is used as a blank, to set the spectrophotometer reading for 0 absorbance (100% transmittance). If you have done Experiment 27, you will recall that the spectrophotometer must be adjusted to 0 and 100% transmittances before meaningful readings can be taken. The patient is then injected with a harmless, physiologically useless colored dye. The purpose of this is to test the liver, to measure the time needed to clear this foreign material from the bloodstream. The dye, bromsulphalein (sulfobromophthalein), works nicely for this purpose and is quantitatively injected into a patient's bloodstream in a calculated amount of 5 mg/kg of body weight.

Immediately after the bromsulphalein injection, a second 10 mL blood specimen is taken from another part of the patient's body, and its fibrinogen and formed elements are removed (Experiment 33). On the assumption that none of the dye has yet been cleared, this serum specimen is used as the blank to establish the maximum absorbance (minimum percent transmittance) of the contaminated blood. Incidentally, for the bromsulphalein system the maximum absorption occurs at 575 nm, so of course the spectrophotometer is set for that particular wavelength.

Exactly 45 minutes after the dye injection a third 10 mL blood sample is taken from the patient, and the serum from this specimen is measured for absorbance and percent transmittance. If the dye concentration is now less than 5% of the maximum concentration of 45 minutes earlier, the liver is functioning normally.

MATERIALS NEEDED

Obtain three colorimeter cuvettes, a 50 mL volumetric flask, and a 10 mL sample of a simulated blood serum specimen for analysis. This unknown represents a serum specimen drawn from a patient 45 minutes after injection with a calculated amount of bromsulphalein.

EXPERIMENTAL PROCEDURE

A standard reference curve must be constructed in order to establish the bromsulphalein concentration in your unknown. This will be a joint effort by the entire class. You will contribute one data point to the curve. You are expected to use your reading and the readings contributed by classmates to construct, on the grid provided on the data sheet, a standard reference curve.

A. Preparation of Standard Reference Solution

Refer to Table 35.1 and prepare a standard reference solution in accordance with the initial of your surname. For example, if you are a Kelly, obtain in a

TABLE 35.1

Initial of Surname	Milliliters of Standard Bromsulphalein (100 mg/liter)	Concentration of Reference Standard (mg/liter)
A–C	5	10
D–F	10	20
G–I	15	30
J–L	20	40
M–O	25	50
P–R	30	60
S–U	35	70
V–X	40	80
Y–Z	45	90

50 mL volumetric flask 20 mL of the standard solution of bromsulphalein from the automatic dispensing buret, then dilute this with water to the 50 mL mark. Mix thoroughly to get a standard solution containing 40 mg of bromsulphalein per liter of solution.

B. Adjustment of the Spectrophotometer

As described in more detail for Experiment 27, set the instrument wavelength at 575 nm; turn on the instrument (if not already done) and allow 15 minutes for thermal equilibration.

In a clean spectrophotometer cuvette, mix exactly 4.50 mL of 0.05 N NaOH and 0.50 mL of water from the dispensing burets provided in the lab. This is your blank solution. (In a clinical setting, 0.50 mL of preinjection serum would be used instead of the water.) *Mix well* and wipe the outside of the cuvette clean and dry.

As described in Experiment 27, set the dial for 0% transmittance with the left front knob (on a Spectronic 20) with the cuvette holder empty and closed. Insert the cuvette containing your blank solution, close the cover, and set the percent transmittance to 100 using the right front knob. (Be sure that the etched mark on the cuvette is aligned with the arrow on the cuvette holder.) Remove the cuvette, close the cover, and reset at 0% transmittance if it does not return to 0 automatically. Reinsert the blank and make sure that the dial reading is at 100% transmittance. Save this cuvette of blank for subsequent use.

C. Determination of Absorbance of Standard Solution

Into a second clean cuvette, pipet exactly 0.50 mL of your standard solution and 4.50 mL of 0.05 N NaOH from the dispenser. Mix well and double-check your instrument setting with the blank. Determine and record the absorbance as well as the percent transmittance.

D. Plotting and Reporting Reference Point

Two responsibilities are yours at this point. On the grid system posted in the lab, plot your experimentally determined absorbance vs. the reference concentration, and initial this point. Secondly, on the chart also posted, record your name, the concentration of your reference solution, and the absorbance found. You will be provided a copy of this chart for use in the construction of your own standard reference curve near the end of the lab period.

E. Determination of Absorbance at Time of Injection

The standard bromsulphalein in the dispenser from which you prepared your reference standard has been prepared to approximate the serum concentration immediately after a bromsulphalein injection. Mix 0.50 mL of this standard with 4.50 mL of 0.05 N NaOH in your third cuvette and mix well. Reset your spectrophotometer to 0 and 100% transmittance and then determine the absorbance of the solution you just prepared. Record.

F. Determination of Absorbance of Unknown Specimen

Empty, rinse, and dry cuvettes 2 and 3. Into one of these cuvettes mix exactly 0.50 mL (pipet) of your unknown specimen and 4.50 mL of 0.05 N NaOH. (Remember that this unknown solution simulates the specimen drawn from a patient after 45 minutes.) Mix well. Reset the spectrophotometer to 0 and 100% transmittance, and determine the absorbance of this unknown. Repeat this entire part F a second time using the other cuvette and record your findings.

G. Plotting of Standard Reference Curve

Using the chart of absorptions for reference compounds posted in the lab (or duplicated and distributed by your instructor), plot a standard reference curve on the grid provided on the data sheet and draw the best *straight* line to accommodate these points. Determine the slope of this line (refer to Figure 27.1).

H. Calculation of the Liver Function and Diagnosis

From your reference curve and its slope, determine the concentration in mg/liter of the bromsulphalein for your hypothetical patient at injection time and after 45 minutes, and from this calculate the percent retention of bromsulphalein after 45 minutes. Record your findings on the data sheet.

Clean up your cuvettes, volumetric flask, and anything else checked out and return them to the storeroom.

DATA SHEET

Name: ______________________ Lab Section: ______ Date: ____________

BROMSULPHALEIN ANALYSIS: A LIVER FUNCTION TEST

C. Absorbance of your assigned standard solution: __________

Percent transmittance: __________

E. Absorbance of unknown serum specimen at time of injection: __________

F. Absorbance of unknown serum specimen 45 min after injection:
__________ (first reading)

__________ (second reading)

H. Unknown number __________ has a concentration of: __________ mg/liter

The percent retention of the bromsulphalein was:

On the basis of the simulated tests, what would you diagnose regarding the patient whose tests you ran?

Standard Reference Curve (from data cooperatively compiled by the class)

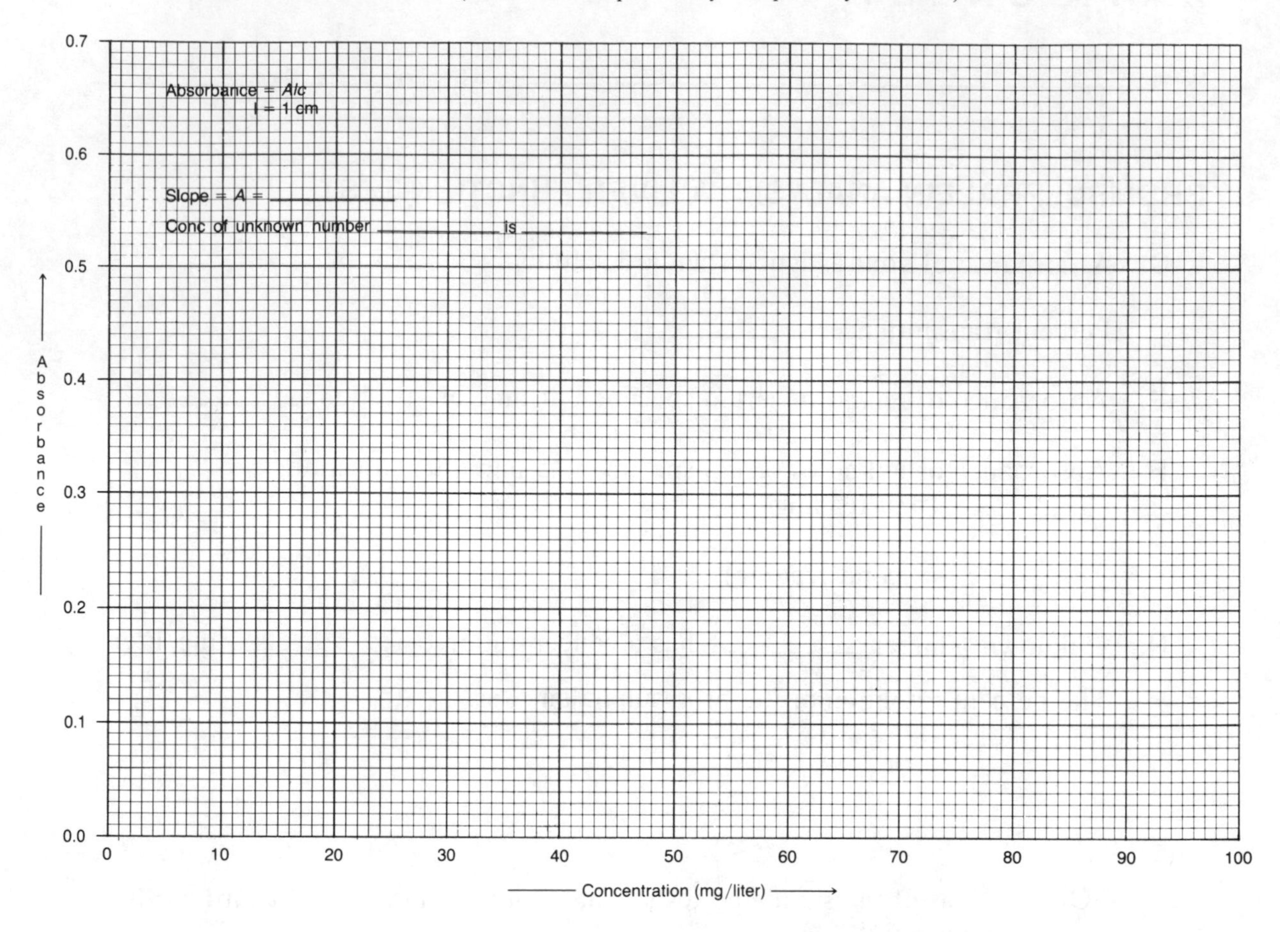

QUESTIONS

1. The Spectronic 20 contains prisms (or gratings) that make it possible to resolve white light into any wavelength component between 375 and 650 mμ (nm). Suppose you were designing a spectrophotometric analysis for a blue-green solute but did not know what to use for the maximum absorbance wavelength. (In this experiment we were given the value of 575.) How could you go about finding the maximum absorbance wavelength for the proposed system? Briefly outline what you would do.

2. The blood serum of a patient subjected to the bromsulphalein liver function test showed an absorption of 0.29 after 45 minutes. Assuming your reference curve (i.e., that of the class) to be correct, what was the bromsulphalein concentration of the blood and what was the percent retention of the dye?

EXPERIMENT 36

DDTC Analysis (Spectrophotometric)

OBJECTIVES

1. To gain an expanded proficiency in extraction techniques.
2. To gain an expanded proficiency in spectrophotometric analyses.
3. To quantitate DDTC in aqueous solution.

PRELABORATORY QUESTIONS

1. On the basis of the structures of DDTC and of $Cu(DDTC)_2$ shown below, explain why the former is so much more water soluble than the latter.
2. DDTC nicely forms complexes with polyvalent metal ions. Write a balanced equation for the reaction between DDTC and $CdCl_2$.
3. In this experiment, DDTC is converted to $Cu(DDTC)_2$ before it is analyzed spectrophotometrically. Give *two* reasons why the DDTC is not analyzed directly.
4. Convert 100 μg/mL to mg/liter. To ppm.

DISCUSSION

Diethyldithiocarbamate, DDTC, is a versatile compound with a rapidly growing list of medicinal and industrial uses. Clinically, it is used to reduce dose-limiting toxicity of antitumor drugs as well as to treat chronic alcoholism. It is especially useful as a rescue drug for nickel carbonyl poisoning. Industrial applications include the cross-linking of rubber and the quantitative analysis of copper. For these reasons a simple, rapid technique for monitoring its concentration is of great value. That is what this experiment is all about.

We will capitalize on the unique ability of DDTC to form complexes with polyvalent metal ions. Specifically, we will make the yellow Cu^{2+} complex since it is easy to analyze with the spectrophotometer. By treating a solution of DDTC with an excess of Cu^{2+} ions, all the DDTC is converted to a relatively nonpolar and lipophilic $Cu(DDTC)_2$ complex. This complex is nicely extracted on a lipophilic surface (e.g., a C-18 SEP-PAK), washed free of water-soluble impurities, and eluted with an appropriate organic solvent such as methanol. The colored methanol solution can then be quantitated with a spectrophotometer set at 430 nm.

Water-soluble DDTC reacts quantitatively with Cu^{2+}:

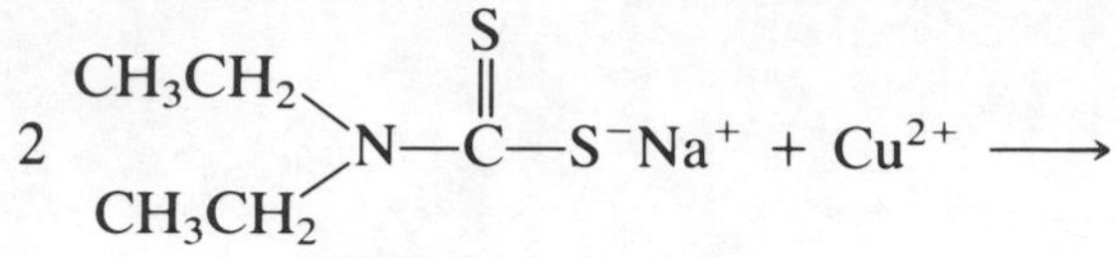

$$2\ (CH_3CH_2)_2N{-}\overset{\overset{\large S}{\|}}{C}{-}S^-Na^+ + Cu^{2+} \longrightarrow$$

DDTC (colorless)
Sodium diethyldithiocarbamate
(ionic—hydrophilic)

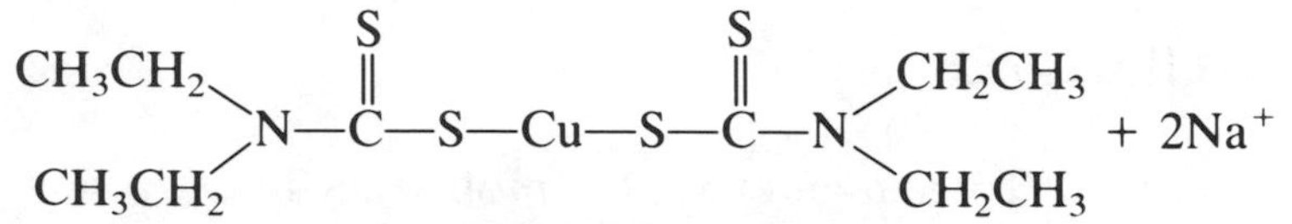

$$(CH_3CH_2)_2N{-}\overset{\overset{\large S}{\|}}{C}{-}S{-}Cu{-}S{-}\overset{\overset{\large S}{\|}}{C}{-}N(CH_2CH_3)_2 + 2Na^+$$

$Cu(DDTC)_2$ (yellow)
Cupric diethyldithiocarbamate
(nonionic—lipophilic)

In preparation for this analysis it is suggested that you review Experiment 21 to refresh your memory of column extraction techniques and Experiment 27 for a review of spectrophotometry.

In Experiment 27 you constructed a standard reference curve. In Experiment 35 you collaborated with classmates to do the same. In this experiment the curve is constructed exclusively from your own data and is the most time-consuming part of the analysis.

MATERIALS NEEDED

Obtain a preprepared unknown DDTC solution, three spectrophotometer cuvettes, a C-18 SEP-PAK, a graduated 5 mL pipet, a syringe for use with the extraction column (SEP-PAK), a filter flask equipped with a one-hole stopper, six 150 mm test tubes, and twelve 10 mL test tubes. All other necessary materials and reagents will be provided in the laboratory in suitable dispensers.

EXPERIMENTAL PROCEDURE

A. Construction of a Standard Curve

1. *Preparing reference DDTC solutions:* Number six clean 150 mm test tubes. Into tube 1 add 10.0 mL of the standard 200 μg/mL DDTC

solution from the automatic buret assembly provided. Into tubes 2 through 6 add 5.0 mL of 0.20 M Na_2CO_3 solution. An automatic buret assembly is provided.

With a pipet, transfer 5.0 mL from tube 1 into tube 2 and mix by gently stirring with the tip of the pipet.

Pipet 5.0 mL from tube 2 to tube 3 and mix with the pipet tip.

Pipet 5.0 mL from tube 3 to tube 4 and mix as before. Then transfer 5.0 mL from tube 4 to tube 5 and mix, and finally transfer 5.0 mL from tube 5 to tube 6 and mix.

You will then have the following sequence of standards:

Tube 1	Tube 2	Tube 3	Tube 4	Tube 5	Tube 6
200 μg/mL (5 mL)	100 μg/mL (5 mL)	50 μg/mL (5 mL)	25 μg/mL (5 mL)	12.5 μg/mL (5 mL)	6.25 μg/mL (10 mL)

2. *Developing the colored $Cu(DDTC)_2$ complex standards:* Number twelve 10 mL test tubes and arrange them by twos as indicated in the chart below. You are preparing duplicate color complexes for the construction of your standard reference curve. To each tube add, in order as indicated, 5.0 mL of distilled water, 1.0 mL of the appropriate standard reference solution from step 1, and finally 2.0 mL of the copper citrate buffer solution. Mix thoroughly after each addition of reference standard and of buffer. Automatic dispensers are provided for the water and buffer solution, but use a 1 mL ($\frac{1}{100}$) pipet for additions of your standard reference solutions. Have a small beaker of distilled water available to rinse your pipet between additions from the different standards. The colored $Cu(DDTC)_2$ complex is formed within seconds, so you can proceed directly to step 3.

Tube #'s	**Water Added**	**Std. Reference Added**	**Cu Citrate Buffer Added**
1 and 2	5.0 mL	1.0 mL (200 μg/mL)	2.0 mL
3 and 4	5.0 mL	1.0 mL (100 μg/mL)	2.0 mL
5 and 6	5.0 mL	1.0 mL (50 μg/mL)	2.0 mL
7 and 8	5.0 mL	1.0 mL (25 μg/mL)	2.0 mL
9 and 10	5.0 mL	1.0 mL (12.5 μg/mL)	2.0 mL
11 and 12	5.0 mL	1.0 mL (6.25 μg/mL)	2.0 mL

Obtain about 40 mL of dry methanol in a clean, dry Erlenmeyer flask. Have in readiness a pipet for measuring out 3.0 mL portions of this methanol.

3. *Extracting the $Cu(DDTC)_2$ complex:* Place the long end of a C-18 SEP-PAK column on the Luer tip of your syringe. If you did Experiment 21 or 27, you should have this SEP-PAK in your lab drawer. Prime the column by adding 3 mL of methanol to the syringe and *slowly* forcing it through the column into the sink.

Rinse the primed column by slowly flushing with 5 mL of distilled water. The column is now ready for use.

Transfer the entire contents of tube 1 from step 2 into the syringe. Slowly force this solution (one to two drops/second) through the column and into the sink. The yellow-colored $Cu(DDTC)_2$ is now adsorbed in the column.

Rinse the adsorbed complex free of adhering ions by slowly passing 5 mL of distilled water through the column.

Dry the column and its adsorbed $Cu(DDTC)_2$ by aspirating for at least 2 minutes on a stoppered filter flask (see Figure 21.3). *Good aspiration is essential!* It is most helpful first to force air rapidly through the column several times with the syringe, and then to detach the column from the syringe and tap sharply on a dry towel to dislodge water droplets from the column stem. *Then*, aspirate for 2 minutes. While aspirating, towel dry the syringe barrel and plunger and the spectrophotometer cuvette to be used.

4. *Eluting the $Cu(DDTC)_2$ complex:* Add exactly 3.0 mL of dry methanol to the *dried* syringe and slowly elute the adsorbed and *dried* $Cu(DDTC)_2$ into a clean, *dry* spectrophotometer cuvette. (*The presence of water greatly interferes with the elution process and also causes a turbidity in the eluate that invalidates colorimetric analysis.*)

5. *Determining the absorbance of the eluate:* Set the spectrophotometer for 430 nm and be sure that it has had time to reach thermal equilibrium (switch on for 15 minutes). Add about 3 mL of dry methanol to a cuvette and save this for your blank. Set the spectrophotometer for 0% transmittance (left front knob on the Spectronic 20) with the cuvette holder closed and empty. Insert the reagent blank (cuvette of methanol) and set for 100% transmittance with the right front knob. Repeat the 0 and 100% transmittance settings. Determine the absorbance and percent transmittance of the eluate from step 4. Record your findings.

Repeat steps 3 through 5 for each of the remaining eleven standards, recording all readings as taken. *You need not repeat the priming and rinsing preliminaries done at the start of step 3.* Simply transfer *all* of tube 2 to the syringe and proceed with the extraction, washing, drying, elution, and the determination of absorbance as was done with tube 1. Repeat in the same sequence for the remaining ten tubes. It is important that absorbance readings be made as soon as possible after the elution to minimize a troublesome turbidity that develops with time. As already mentioned, studiously avoid any possible contamination with water after the drying process!

6. *Constructing the standard reference curve:* Using the grid on the data sheet, plot the averaged absorbances on the x-axis vs. DDTC concentration (μg/mL) on the x-axis and draw the *best straight line* through the points. (Remember that percent transmittance readings are the most reliable in the 25–75% range.)

Determine the slope (see Figure 27.1) of the line. This enables you to calculate the concentration of DDTC in your unknown by use of the equation:

$$\text{concentration} = \text{absorbance}/\text{slope}$$

B. Analysis of Unknown DDTC Solution

Your unknown DDTC solution should be developed, isolated, and checked for absorption of light in exactly the same way as described for the reference solutions. Simply add 1.00 mL of the unknown to 5.00 mL of distilled water, buffer with 2.00 mL of the copper citrate buffer, and mix. Transfer *all* of this solution to the syringe, absorb the $Cu(DDTC)_2$ on the column, rinse with 5 mL of water, dry *thoroughly* on the stoppered filter flask, and finally elute with 3.0 mL of methanol.

In other words, repeat steps 2 through 4 under part A, substituting the unknown solution for the standard reference solution.

Determine the absorbance of this eluate as before, after again checking the spectrophotometer for 0 and 100% transmittances with the methanol blank. Double-check to make sure that the wavelength setting of the instrument remains at 430 nm.

Run two more trials (steps 2 through 4) on your unknown and average the resulting absorbances.

From the slope of your standard curve calculate the concentration of DDTC in your unknown in units of $\mu g/mL$. Record all results on your data sheet.

Rinse your SEP-PAK with 3 mL of methanol; you may need it for use in Experiment 37. Clean and dry the cuvettes, syringe, pipets, and suction flask and check them back in.

3. Analysis of Unknown DDT Solution

[illegible]

DATA SHEET

Name: ______________________ Lab Section: ______ Date: ____________

DDTC ANALYSIS (SPECTROPHOTOMETRIC)

Unknown number: ____________

A. Data for Standard Reference Curve

Wavelength setting: ____________

Concentrated DDTC ($\mu g/mL$)	Tube No.	Absorbance	Average Absorbance
200	1	________	____________
	2	________	
100	3	________	____________
	4	________	
50	5	________	____________
	6	________	
25	7	________	____________
	8	________	
12.5	9	________	____________
	10	________	
6.25	11	________	____________
	12	________	

Slope: $\left(\frac{\text{absorbance}}{\mu g/mL}\right)$ = ____________

Unknown: __________

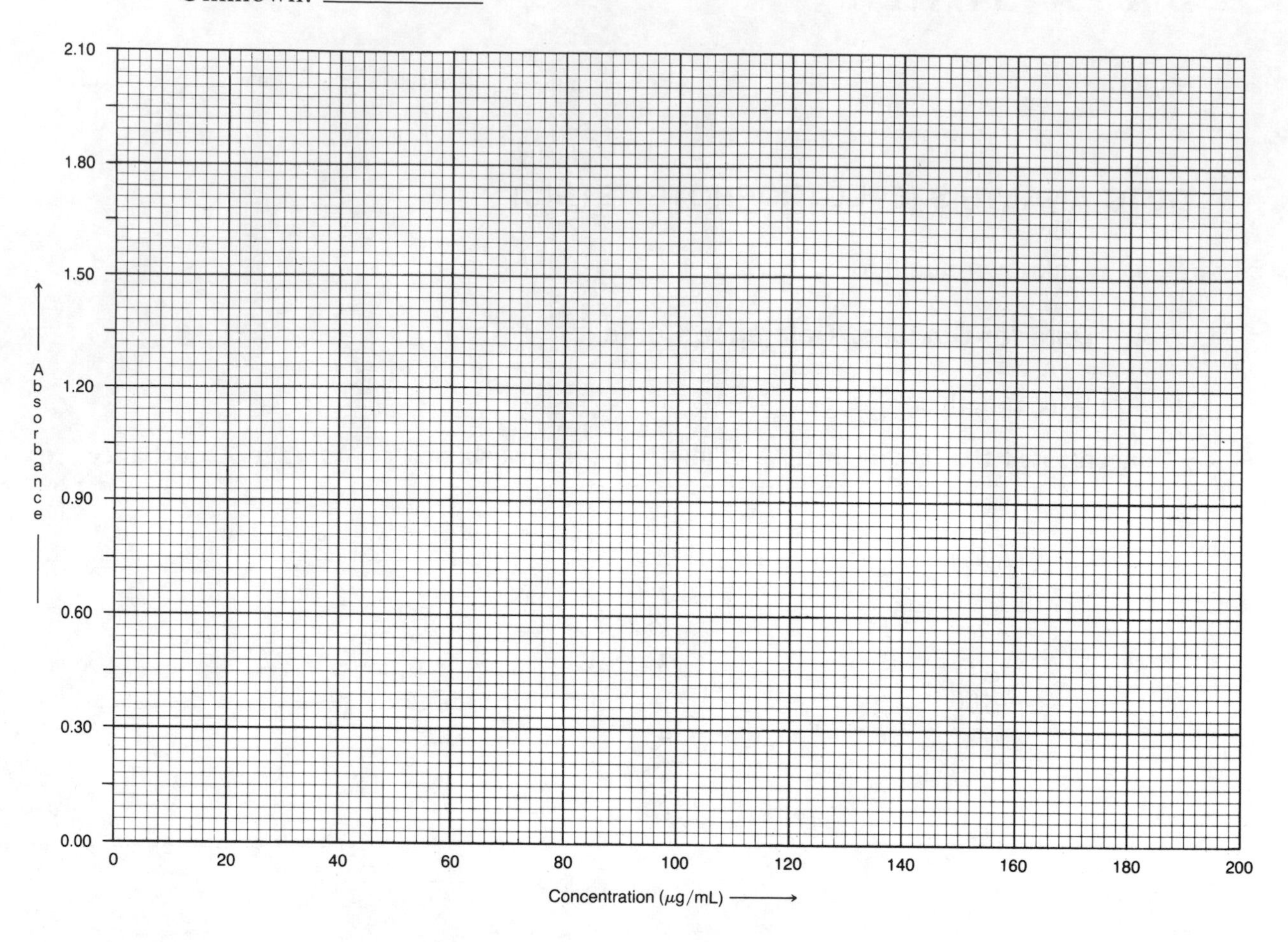

B. Unknown Analysis

Trial Number	Absorbance	Average Absorbance	Concentrated DDTC (μg/mL)
1	__________		
2	__________	__________	__________
3	__________		

QUESTIONS

1. The hydrophobic absorbant in the C-18 SEP-PAK is designated by the trivial name octadecyl. What would be the structure of octadecyl? Is it indeed a chemical compound?

2. What is Beer's law, and what is meant by Beer's law conformity? (Define all terms.)

3. Why is it necessary to dry the absorption column (SEP-PAK) with a stream of air before proceeding with elution?

4. Polycyclic aromatic hydrocarbons (PAH) can be found in the surface waters in areas where insecticides are used. These compounds are nonpolar and highly resistant to biodegradation. Describe a procedure by means of which you might concentrate the PAH in river water such that analyses of these PAH compounds might be made.

EXPERIMENT 37

The Determination of the Half-life of Diethyldithiocarbamate (DDTC)

OBJECTIVES

1. To gain experience in experimental chemical kinetics measurements.
2. To understand how quenching can be applied to the study of reaction rates.
3. To establish the half-life of a medicinally useful compound.

PRELABORATORY QUESTIONS

1. Define half-life.
2. In this experiment we study the stability of DDTC at a pH of 6.0. What is the hydrogen ion concentration in our reaction?
3. What is meant by the expression "quenching a reaction"?
4. Carbamic acids are not uncommon in metabolic reaction sequences where they readily decarboxylate to give amines. Referring to the structure for diethyldithiocarbamic acid provided in the following discussion, draw the structure for *diethylcarbamic acid* (in which the two sulfurs are replaced by oxygens) and write the equation for its spontaneous decarboxylation.

DISCUSSION

The versatile complexing agent diethyldithiocarbamate was introduced in Experiment 36. This experiment will extend that study. Medicinal application of DDTC as it is used for the reduction of the toxic effects of platinum antitumor drugs or of nickel carbonyl requires intravenous injection. The question then arises: How stable is DDTC when in the bloodstream? What is its active life in a living organism?

It was learned early on that DDTC could not be administered orally. The acid conditions of the stomach cause its decomposition to carbon disulfide and diethylammonium ion:

$$\underset{\substack{\text{DDTC}\\ \text{(stable, pH > 8)}}}{(CH_3CH_2)_2N\overset{S}{\overset{\|}{C}}—S^-Na^+} \underset{}{\overset{H^+}{\rightleftharpoons}} \underset{\substack{\text{Diethyldithiocarbamic}\\ \text{acid}\\ \text{(unstable, pH < 8)}}}{\left[(CH_3CH_2)_2N\overset{S}{\overset{\|}{C}}—SH\right]} \xrightarrow{-CS_2} (CH_3CH_2)_2NH \xrightarrow{H^+} \underset{\text{Diethylammonium ion}}{(CH_3CH_2)_2NH_2^+}$$

Even near neutrality this decomposition, though slow, goes on. It is a function of the hydrogen ion concentration that is low but real even in slightly alkaline blood. The in vivo life of DDTC is definitely limited. How limited is it?

In this experiment the half-life of DDTC at pH 6.0 is determined for room temperature. The half-life is defined as the time required for half of a freshly prepared DDTC solution to decompose. Our experimental findings should be a conservative approximation of the half-life of DDTC in blood. In blood, the pH is about 7.4 (slightly alkaline) and the temperature is 37° (warmer than room temperature). The high pH should cause a slower reaction, whereas the higher temperature would speed a reaction. At any rate, we will monitor the rate of disappearance of DDTC by establishing the concentration of DDTC remaining at regular time intervals. We will accomplish this by quenching (stopping abruptly) the reaction in samples at timed intervals and then measuring spectrophotometrically the amount of DDTC left in the sample taken (the aliquot). The quenching is done by adding a buffered copper citrate solution that precipitates any existing DDTC as a stable $Cu(DDTC)_2$ complex. The reaction is the same as that practiced in Experiment 36:

$$2\underset{\substack{\text{DDTC (colorless)}\\ \text{(hydrophilic)}}}{(CH_3CH_2)_2N\overset{S}{\overset{\|}{C}}—S^-} + Cu^{2+} \longrightarrow \underset{\substack{\text{Cu(DDTC)}_2\text{ (yellow)}\\ \text{(lipophilic)}}}{(CH_3CH_2)_2N\overset{S}{\overset{\|}{C}}—S—Cu—S—\overset{S}{\overset{\|}{C}}N(CH_2CH_3)_2}$$

The yellow copper complex is adsorbed on a column, washed, dried, eluted, and spectrophotometrically analyzed (as in Experiment 36).

Your instructor will inform you as to whether or not you might work with a partner.

MATERIALS NEEDED

Two colorimeter cuvettes, a Luer tip syringe, a C-18 SEP-PAK, a 1 mL volumetric pipet, and twenty-one 10 mL test tubes are needed. All other reagents and materials will be provided in the laboratory.

EXPERIMENTAL PROCEDURE

A. Preparation of Materials Used

It is most important that contamination by copper does not occur. As Experiment 36 demonstrated, cupric ion will precipitate DDTC from aqueous solutions as a stable complex. If you did Experiment 36, copper contamination is highly probable. Wash and rinse thoroughly all glassware to be used, especially the syringe, your cuvettes, and your test tubes.

Prepare 21 test tubes with labels T_i and T_1 through T_{20}. Clean and rinse a 1 mL pipet graduated to $\frac{1}{100}$ mL. As this experiment proceeds, a flocculant brown precipitate betrays the presence of contaminating copper. You don't want to see *this* reaction until you quench by adding copper citrate buffer.

Prime your SEP-PAK by slowly forcing 5 mL of methanol through it. (*If your SEP-PAK was used in Experiment 36*, next remove contaminating copper and iron carbonates and hydroxides by slowly purging the column with 5 mL of 0.1 N HCl.) Finally, rinse with 10 mL of distilled (or deionized) water. Your column is then ready for use.

B. Preparation of Buffered DDTC

Transfer 29 mL of phosphate buffer (pH 6.0) into a 50 mL Erlenmeyer flask. To this add 1.0 mL of the 0.025 M DDTC provided in the dispensing buret. Record the time, t_i, and temperature of the solution while mixing **carefully** with your thermometer. The decomposition reaction is now underway, and you should be prepared to do the next part as quickly as possible.

C. Sampling and Quenching

Pipet 1.0 mL of the reaction solution into tube T_i. To this add 3.0 mL of distilled water (dispenser) and 1.0 mL of the copper citrate buffer solution provided in an automatic dispenser. Note the exact time and mix. You have just quenched the reaction. The $Cu(DDTC)_2$ formed represents, mole for mole, the DDTC remaining as of the moment the buffer was added. The system is now stabilized and can be quantitated at your convenience.

Five minutes later repeat this sampling in tube 1. That is, pipet 1.0 mL of your reaction mixture into 3 mL of water in tube 1, and then, noting the time, mix in 1.0 mL of copper citrate buffer.

Repeat this sampling process every five minutes for the next eight tubes and then at ten minute intervals for tubes numbered 11 through 20. Be sure to record all times accurately. If the time interval between two samplings was 6 minutes instead of 5 minutes, it must be recorded as such. As this routine is established, you may save time by proceeding with the quantitation of these quenched aliquots between samplings.

D. Isolation of $Cu(DDTC)_2$

As was done in Experiment 36, pour the contents of tube T_i into the syringe—with the primed and rinsed SEP-PAK attached—and elute. To ensure a quantitative transfer, add 2 mL of water to tube T_i, slosh it around, and add this wash also to the syringe.

Extract the $Cu(DDTC)_2$ by slowly forcing the solution through the column and into the sink. Wash with 5 mL of distilled water. Dry thoroughly. To expedite drying, force air rapidly through the column with the syringe plunger several times, holding the syringe with column aimed at the floor. Then disconnect the column and tap sharply on a dry towel to dislodge any water droplets retained in the stem. Finally, aspirate for 2 minutes on a stoppered filter flask, as illustrated in Figure 21.3. If you detect liquid droplets in the column stem during aspiration, interrupt the suction and tap on dry toweling to dislodge them. While aspirating, thoroughly towel dry the syringe barrel and plunger and the cuvette to be used as the receptacle for the next elution step.

Elute the yellow $Cu(DDTC)_2$ from your dried syringe directly into a dry cuvette with a 3.0 mL of dry methanol.

E. Quantitation of Quenched Aliquot

If the spectrophotometer is not already warmed up, turn it on and allow 15 minutes for thermal equilibration. Set the wavelength at 430 nm. With the cuvette holder empty and closed, set the instrument for 0% transmittance with the left front knob. Insert a clean cuvette containing 3 mL of methanol (etched line on the forward mark) and, with cover closed, set the percent transmittance at 100 using the right front knob. Repeat these two settings.

Insert the cuvette of eluate and record the absorbance and percent transmittance. (*If cloudiness is evident, absorbance readings are unreliable. You failed to exclude moisture, and you must in future runs pay greater attention to the drying.*)

Repeat the sequence (parts D and E) for each of the remaining 20 timed aliquots, making certain that all times and absorbances are correctly recorded.

F. Data Analysis

From the data recorded on the data sheet and on the grid provided, plot the absorbance (x-axis) vs. the reaction time in minutes (y-axis). Draw the *best straight line* through these points, bearing in mind that the spectrophotometer readings are the most reliable between 25% and 75% transmittance.

If, as we expect and hopefully have shown, this system exhibits Beer's law conformance, the half-life, $t_{1/2}$, must be that time at which absorbance is half of what it was at t_i (or half of what it was at *any* arbitrarily selected starting time).

Calculate the half-life of the DDTC starting at t_i and from at least two other starting points of your own selection. Average these half-lives and complete your data sheet.

Clean and dry cuvettes, syringes, and other materials checked from the storeroom and return them.

DATA SHEET

Name: ______________________ Lab Section: ______ Date: __________

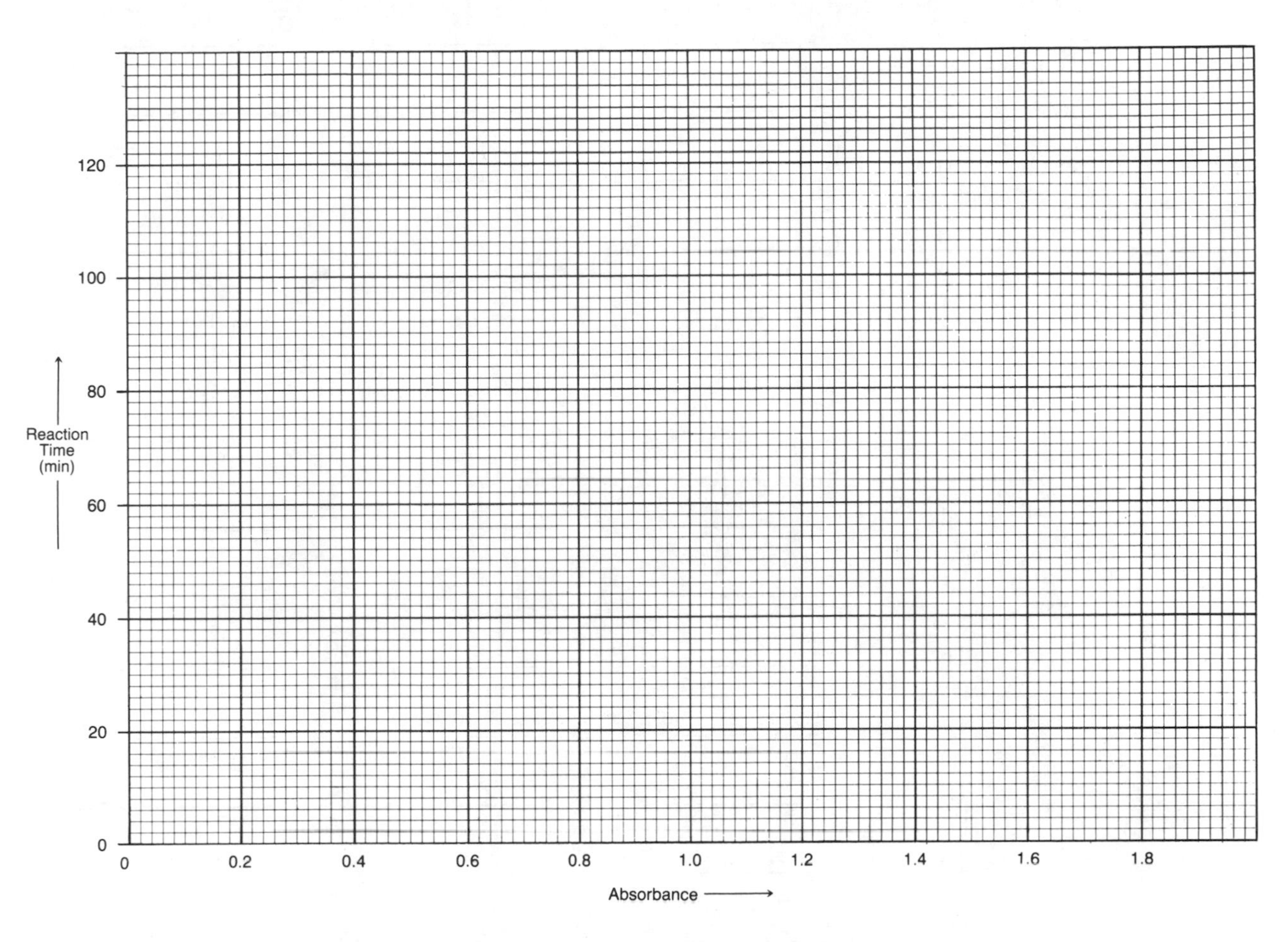

THE DETERMINATION OF THE HALF-LIFE OF DDTC

Temperature of solution	pH of solution	[DDTC] in μg/mL
______	______	______

Data

Tube #	Rx time (min)	% Transmittance	Absorbance
T_i	______	______	______
1	______	______	______
2	______	______	______
3	______	______	______
4	______	______	______

5			
6			
7			
8			
9			
10			
11			
12			
13			
14			
15			
16			
17			
18			
19			
20			

Half-life calculations:

1. From T_i:

2. From Tube # __________:

3. From Tube # __________:

Average half-life of DDTC (in minutes): __________

Sources of error in the investigation:

QUESTIONS

1. DDTC generally decomposes much more rapidly in human urine than it does in human plasma. Why is this to be expected?

2. A student performed this same experiment using a 0.06 M phosphate buffer as the medium instead of the 0.6 M phosphate buffer that we used. The results obtained were reasonable at the beginning of the experiment, but the half-lives got longer and longer as the experiment continued. Refer to the decomposition equation in the discussion section and explain why this happened.

3. One way that the liver can remove DDTC from the blood is to complex the DDTC with glucuronic acid and excrete the complex in the urine. A researcher once tested for the presence of DDTC in urine by adding a few drops of 1 M $CuSO_4$. He got a clear colorless solution indicating an absence of DDTC in the specimen. However, when he checked the same tube the following day, he discovered that a brown precipitate had formed—as one would expect if DDTC had been introduced. He dissolved this precipitate in methanol and obtained a yellow solution with an absorption maximum at 430 nm, just as would be expected for $Cu(DDTC)_2$. Using equations, describe what had happened.

A 8
B 9
C 0
D 1
E 2
F 3
G 4
H 5
I 6
J 7